Geological Society of America
Memoir 188

Palynological Correlation of Major Pennsylvanian (Middle and Upper Carboniferous) Chronostratigraphic Boundaries in the Illinois and Other Coal Basins

Russel A. Peppers
Coal Section
Illinois State Geological Survey
Natural Resources Building
615 East Peabody Drive
Champaign, Illinois 61820-6964

1996

Published by The Geological Society of America, Inc.
3300 Penrose Place, P.O. Box 9140, Boulder, Colorado 80301

Printed in U.S.A.

GSA Books Science Editor Abhijit Basu

Library of Congress Cataloging-in-Publication Data
Peppers, Russel A. (Russel Allen), 1932-
    Palynological correlation of major Pennsylvanian (Middle and Upper
Carboniferous) chronostratigraphic boundaries in the Illinois and
other coal basins / Russel A. Peppers.
       p. cm. -- (Memoir ; 188)
    Includes bibliographical references and index.
    ISBN 0-8137-1188-6
    1. Geology, Stratigraphic--Pennsylvanian. 2. Palynology.
3. Stratigraphic correlation. 4. Coal mines and mining. I. Title.
II. Series: Memoir (Geological Society of America) ; 188.
QE673.P465   1996
551.7'52'09773--dc20   95-52312
                             CIP

10 9 8 7 6 5 4 3 2 1

Contents

Geological Society of America
Memoir 188
1996

Palynological Correlation of Major Pennsylvanian (Middle and Upper Carboniferous) Chronostratigraphic Boundaries in the Illinois and Other Coal Basins

ABSTRACT

Palynology provides chronostratigraphic correlations of Pennsylvanian (Middle and Upper Carboniferous) strata between the Illinois basin, which is the main focus of this study, and other coal basins in the United States, western Europe, and the Donets basin. Published reports and recent palynological analyses of coal samples from the Illinois basin, western part of the Midcontinent, and the Appalachian coal region were used in making the correlations. Correlations are made on the basis of first and last appearances and major changes in relative abundance of spore taxa. Studies of plant megafossils and faunas are discussed briefly in relation to some of the palynological correlations.

Stratigraphers in the Midcontinent have considered the Lower Pennsylvanian Series equivalent to the Morrowan Stage, the Middle Pennsylvanian equivalent to the Atokan and Desmoinesian Stages, and the Upper Pennsylvanian equivalent to the Missourian and Virgilian Stages. I propose extending this nomenclature to the Illinois Basin.

The chronostratigraphic classification of the Pennsylvanian System in the western part of the Midcontinent region was adopted by the Illinois State Geological Survey for the Illinois basin largely on the basis of studies of marine fossils. The precise position of the Morrowan–Atokan Stage boundary in Illinois is still not satisfactorily known. It generally has been correlated with the base of the Tradewater Formation but is herein correlated with the top of the Reynoldsburg Coal Bed, which is at the base of the formation. The top of the Morrowan occurs just above the base of the Tradewater Formation in western Kentucky and in the upper part of the Mansfield Formation in Indiana. I propose that the Atokan–Desmoinesian Stage boundary in Illinois correlates with the top of the Seville Limestone Member and equivalent limestone beds in Indiana and western Kentucky.

The Desmoinesian–Missourian (Middle–Upper Pennsylvanian) boundary is correlated with the base of the Lake Creek Coal Member or, if that is not present, at the top of the Lonsdale Limestone Member in western Illinois and the top of the middle bench of the West Franklin Limestone Member in eastern Illinois and western Indiana. The Missourian–Virgilian Stage boundary has been correlated, in the absence of adequate palynological data, with strata a little below the Shumway Limestone Member as determined by fusulinid evidence.

The Lower–Middle Pennsylvanian boundary in the Pennsylvanian stratotype in West Virginia proposed by the U.S. Geological Survey (USGS) is older than the Lower–Middle Pennsylvanian boundary of the Midcontinent. The latter correlates with the base of the Pounds Sandstone Member in Illinois and the Bee Spring Sandstone Member in western Kentucky. Although the Middle–Upper Pennsylvanian Series boundary has been correlated with the boundary between the Allegheny and

Peppers, R. A., 1996, Palynological Correlation of Major Pennsylvanian (Middle and Upper Carboniferous) Chronostratigraphic Boundaries in the Illinois and Other Coal Basins: Boulder, Colorado, Geological Society of America Memoir 188.

Conemaugh Formations, the series boundary at the proposed stratotype is at the top of the Charleston Sandstone, which is difficult to correlate outside of its stratotype area. If the Middle–Upper Pennsylvanian boundary of USGS usage is raised slightly to correlate with the Middle–Upper Pennsylvanian boundary in the Midcontinent and the Westphalian–Stephanian boundary in Europe, it would be more easily recognized outside the Appalachian coal region because of the major changes in composition of faunas and coal-swamp floras that occurred at that time.

The Namurian–Westphalian boundary of western Europe is tentatively correlated with the base of the Battery Rock Sandstone Member in Illinois. The Westphalian A–B (Langsettian–Duckmantian) boundary is correlated with the top of the Morrowan, the Westphalian B–C (Duckmantian–Bolsovian) boundary with the lower part of the Tradewater Formation in Illinois, the Westphalian C-D boundary with the Atokan–Desmoinesian boundary, and the Westphalian–Stephanian boundary with the Desmoinesian–Missourian boundary. The lowest stage of the Stephanian (Cantabrian) has not been identified adequately in North America. Correlation of the Westphalian–Stephanian boundary with strata in northern China is possible because of the major palynological changes that occur at the boundary.

The boundary between the Bashkirian and Moscovian Series in the Donets basin is correlated approximately with the top of the lower one-fourth of the Tradewater Formation in Illinois and Kentucky and with the upper one-fourth of the Mansfield Formation in Indiana. The Bashkirian–Moscovian boundary is correlated with the upper part of the Westphalian B. The boundary between the Moscovian and Kasimovian Series is just below the Desmoinesian–Missourian boundary and is approximately equivalent to the Middle–Upper Pennsylvanian boundary as currently designated in the Appalachian coal region.

INTRODUCTION

The purpose of this study is to use existing and new palynological data to outline the correlation of major chronostratigraphic boundaries in the Pennsylvanian System in the Illinois basin, the Western Interior coal province, and the Appalachian coal region as well as in the Middle and Upper Carboniferous in Europe (Plate 1, in pocket). This book is a more detailed documentation and expansion of my previous study (Peppers, 1984), which correlated Pennsylvanian strata in the Illinois basin with Middle and Upper Carboniferous strata in western Europe by comparing palynomorph assemblage zones. In this study, I recommend that several changes in correlation of chronostratigraphic boundaries be considered. Numerous interpretations have been used to define series and stage boundaries, especially where boundaries are poorly defined paleontologically in their type sections and areas. After further investigation by stratigraphers and paleontologists, boundary revisions will certainly be made, through the process of national and international cooperative studies and recommendations. This summary, however, illustrates how palynology can contribute toward an integrated classification and correlation of the Pennsylvanian and Carboniferous strata. It also indicates areas needing the most work to improve current correlations.

The growing number of contributions to our knowledge of the composition of spore assemblages in Pennsylvanian and Carboniferous strata has provided an impetus for comparing palynological successions from one region to another, especially in western Europe (Smith and Butterworth, 1967; Liabeuf et al., 1967; Butterworth, 1969; Coquel, 1976; Coquel et al., 1970, 1976, 1984; Loboziak, 1971, 1972, 1974; Loboziak et al., 1976; Butterworth and Smith, 1976; Van Wijhe and Bless, 1974; Bless et al., 1977; Clayton et al., 1977; Owens et al., 1978; Paproth et al., 1983; Owens, 1984; Van de Laar and Fermont, 1989, 1990).

Palynology of Pennsylvanian coals has been more completely documented in the Illinois basin than it has been elsewhere in North America (Schopf, 1938, Schopf et al., 1944; Kosanke, 1947, 1950, 1964; Guennel, 1952, 1958; Winslow, 1959; Peppers, 1964, 1970, 1979, 1982a, 1982b, 1984, 1993; Phillips et al., 1974, 1985; Phillips and Peppers, 1984; Peppers and Pfefferkorn, 1970; Ravn and Fitzgerald, 1982; Eggert et al., 1983; Mahaffy, 1985). The Illinois State Geological Survey (ISGS) began to investigate the use of palynology for correlating coal beds in the Illinois Basin in the 1930s (McCabe, 1932; Schopf, 1936, 1938; Cady, 1933). Other significant advances in palynology that occurred at the ISGS include the publication of *An Annotated Synopsis of Paleozoic Fossil Spores and Definition of Generic Groups* by Schopf et al. (1944), which produced an early classification of spores and treated them according to the International Code of Botanical Nomenclature. *Pennsylvanian Spores of Illinois and Their Use in Correlation* by R. M. Kosanke (1950) was the first major study to demonstrate that spores could be used to correlate a large number of coal beds, from Early to Late Pennsylvanian, through an entire basin. The history of the development of paleobotany, including palynology, in the Illinois basin was presented by Phillips et al. (1973).

Attempts have been made by palynologists (mostly European) to use spores, principally spore genera, to correlate Upper and Middle Carboniferous strata in Europe with Pennsylvanian strata in the Illinois basin (Potonié and Kremp, 1954, 1956; Hacquebard et al., 1960; Bharadwaj, 1960; Helby, 1966; Butterworth, 1964, 1969; Alpern, 1960; Alpern and Liabeuf, 1969; and Butterworth and Smith, 1976). In 1984, I used unpublished and published palynological data from the Illinois basin to make a more detailed correlation between the Illinois basin and western Europe. Phillips and Peppers (1984) and Phillips et al. (1985) also used, in part, palynological data to correlate Pennsylvanian formations in Illinois with major lithostratigraphic and chronostratigraphic divisions of the Midcontinent, the Appalachian coal region, and Europe.

Major subdivisions of the Pennsylvanian and Middle and Upper Carboniferous

In the Midcontinent of the United States, the Pennsylvanian is divided into the Morrowan, Atokan, Desmoinesian, Missourian, and Virgilian Stages (or, according to some stratigraphers, series). The Morrowan derived its name (Adams, 1904) from the town of Morrow, Washington County, Arkansas, near where the rocks are composed of marine and nonmarine strata. The Atokan was first named (Spivey and Roberts, 1946) for several thousand feet of unfossiliferous sandstone and shale in the Atoka Formation near Atoka, Atoka County, Oklahoma, but no type section was designated. Keyes (1893) named the Desmoinesian Series for deposits along the Des Moines River in Iowa, but a type section for it was also not designated. The Missourian Series was named (Keyes, 1893; Moore, 1931) for mostly limestone strata above the Desmoinesian in Missouri. Moore (1931) named the Virgilian Series for upper Pennsylvanian rocks near Virgil, Kansas.

Major problems in correlating the Morrowan–Atokan boundary still exist because of (1) the lack of diagnostic fossils in the type sections of the Morrow and Atoka Formations in Arkansas and Oklahoma and (2) the disagreement concerning the stratigraphic position of the boundary in the type area. The precise position of the Atokan–Desmoinesian boundary is also in doubt because of the overlap of the upper part of the Atokan in Oklahoma with the lower part of the Desmoinesian in Iowa. A symposium was held in 1982 to discuss the biostratigraphy of the Atokan Series (Sutherland and Manger, 1984); another was held in 1990 to discuss the biostratigraphy of the Morrowan, Atokan, and Desmoinesian in the Midcontinent (Sutherland and Manger, 1992).

The Pennsylvanian was considered a series in some of the early ISGS reports. Worthen (1875) and Weller (1906) divided the "Coal Measures" and the Pennsylvanian into Lower and Upper, respectively, but their usage was dropped soon after 1906. A comparison of stratigraphic ranges of ostracods led Cooper (1946) to use the Midcontinent series names "Ti Valley, Des Moines, Missouri, and Virgil" to subdivide the Pennsylvanian System in Illinois, but this scheme was also not later followed by the ISGS. A lithostratigraphic classification of Pennsylvanian strata and an independent classification of cyclothems in Illinois was presented by Kosanke et al. (1960). A chronostratigraphic classification was not adopted, but correlation was made with the Missouri section, which included the Midcontinent series nomenclature. The names for the Midcontinent series—Morrowan, Atokan, Desmoinesian, Missourian, and Virgilian—were formally adopted by the ISGS on the most recent geologic map of Illinois (Willman et al., 1967), and this classification is the only formal series classification of the Pennsylvanian used by the ISGS. The use of the Midcontinent series classification in Illinois was discussed in more detail by Hopkins and Simon (1975).

Ravn et al. (1984) proposed designating the major lithostratigraphic subdivisions of the Pennsylvanian in Iowa as supergroups—Morrow, Des Moines, Missouri, and Virgil. The equivalent chronostratigraphic units are designated by adding "an" or "ian" as suffixes. The Atokan Series is considered to be the lower part of the Des Moines Supergroup.

The state geological surveys of Kentucky (Williams et al., 1982) and Indiana (Shaver et al., 1986) also use the Midcontinent series classification for the Illinois basin. Shaver et al. (1970) had applied the names Pottsvillian, Alleghenian, Conemaughian, and Monongahelian as series names to indicate correlations with the Appalachian coal region, but these names have been discontinued in the latest compendium of Paleozoic stratigraphy in Indiana (Shaver et al., 1986). Cobb et al. (1981) and Milici et al. (1979) also considered the New River, Kanawha, Charleston Sandstone (Allegheny), Conemaugh, and Monongahela as series in eastern Kentucky and Tennessee, respectively.

The Tri-State Committee on Correlations in the Pennsylvanian System of the Illinois basin was formed by the geological surveys of Illinois, Indiana, and Kentucky to standardize stratigraphic terminology as much as possible between the three states. This has been accomplished for some names of Desmoinesian and Missourian limestones and coal seams (Jacobson et al., 1985). There was also some agreement among the three states with group and formational names (Greb et al., 1992).

The names Lower, Middle, and Upper Pennsylvanian are widely used by stratigraphers in the Midcontinent. Cheney et al. (1945) subdivided the Pennsylvanian in that way, but they did not designate rank. Moore and Thompson (1949) proposed the Ardian Series for the Lower Pennsylvanian, the Oklan Series for the Middle Pennsylvanian, and the Kawvian Series for the Upper Pennsylvanian. The Springeran, Morrowan, Atokan, Desmoinesian, Missourian, and Virgilian were reduced to the rank of stage. The names Ardian, Oklan, and Kawvian fell out of use, and the Morrowan, Atokan, Desmoinesian, Missourian, and Virgilian remained as series names. Zeller et al. (1968), Boardman et al. (1989b), and others (including myself) consider the Morrowan as a stage in the Lower Pennsylvanian Series, the Atokan and Desmoinesian as stages in the Middle Pennsylvanian, and the Missourian and Virgilian as stages in the Upper Pennsylvanian Series.

The names Lower, Middle, and Upper Pennsylvanian Series had been recommended by the Geological Names Committee of the U.S. Geological Survey (USGS) and were adopted in the nomenclature of the USGS (Bradley, 1956). According to Bradley, the names Lower, Middle, and Upper Pennsylvanian Series are to be used in the Appalachian region and in other areas except in the Midcontinent, where the names "Morrow, Atoka, Des Moines, Missouri, and Virgil" are to be used. The Midcontinent region includes Arkansas, Iowa, Kansas, Missouri, Nebraska, and Oklahoma. The Lower Pennsylvanian is supposed to correspond approximately to the Morrowan, the Middle Pennsylvanian to the Atokan and Desmoinesian, and the Upper Pennsylvanian to the Missourian and Virgilian.

In a series of USGS publications by different authors (Professional Papers 1110A–DD) that described comprehensively the Mississippian and Pennsylvanian Systems in each state, most of the authors discussing states east of the Mississippi River used the series names Lower, Middle, and Upper Pennsylvanian according to the guidelines of the USGS. In conjunction with the Ninth International Congress of Carboniferous Stratigraphy and Geology in 1979, a field trip was conducted in parts of Virginia and West Virginia in the area of the proposed Pennsylvanian System stratotype, which was later described by Englund et al. (1986). In the preface of the guidebook, Englund et al. (1979) stated, "In the Appalachian basin, the Pennsylvanian System previously consisted of loosely defined Lower, Middle, and Upper Pennsylvanian Series." Although no reference to Bradley (1956) was given in the guidebook, Bradley's Lower–Middle Pennsylvanian boundary was used. The Middle–Upper boundary was designated, however, as the top of the Charleston Sandstone rather than the top of the Allegheny Formation. Because the USGS uses the names Lower, Middle, and Upper Pennsylvanian Series extensively, correlation of their boundaries with strata in the Illinois basin and elsewhere is of interest. In addition to my own studies, I have used information from the recent palynological studies of coal beds in the stratotype (Kosanke, 1984, 1988a, 1988b, 1988c) to correlate the USGS boundaries with strata in other regions.

The boundaries of the Lower, Middle, and Upper Pennsylvanian proposed by the USGS (Bradley, 1956; Englund et al., 1979, 1986), as determined from stratigraphic studies in the Appalachian region, are not correlative to the Series boundaries previously proposed (Cheney et al., 1945) for the Midcontinent. The Lower–Middle and Middle–Upper Pennsylvanian boundaries in the Midcontinent are younger than the boundaries in the Appalachian region. Important palynological changes occur at the Lower–Middle Pennsylvanian boundary in both regions. However, the Middle–Upper Pennsylvanian boundary as defined in the Midcontinent is palynologically and paleontologically more readily recognizable in North America and Europe than the boundary as defined in the Appalachians. In fact, the Middle–Upper boundary of the Midcontinent, which is correlated with the Westphalian–Stephanian boundary, is marked by the greatest change in coal swamp floras of the Pennsylvanian.

The International Subcommission on Carboniferous Stratigraphy, through its working groups, is investigating geologic sections in the United States and Canada for use as stratotypes for the Pennsylvanian System rather than use the stratotype proposed by the USGS. The geologic sections selected will include fossiliferous marine strata representing, as much as possible, continuous deposition. The study of conodonts, fusulinids, ammonoids, brachiopods, and other marine fossils will be used to correlate strata from the stratotypes to reference sections. Reference sections will include a sufficient number of coal beds and nonmarine strata so that spores and plant compressions can be used to make correlations with regions, such as the Appalachians, that are dominated by nonmarine strata. Economic interest in coal deposits has been a major incentive for studying the Pennsylvanian System; therefore, the need for biostratigraphic control in coal-bearing sequences should also be considered when stratotypes and reference sections are chosen.

The system of classification of Carboniferous rocks in western Europe was instituted at the 1935 International Congress of Carboniferous Stratigraphy and Geology (Jongmans and Gothan, 1937). The classification has since been amended by the Subcommission on Carboniferous Stratigraphy. A significant change was adopted at the Seventh International Carboniferous Conference when the Cantabrian Stage was added to the Stephanian Series. The Cantabrian is thought to represent strata that were eroded from the upper part of the Westphalian D and the lower part of the Stephanian in most of western Europe. The stratotype of the Cantabrian is in the Cantabrian Mountains of northern Spain (Wagner and Winkler Prins, 1985). The European system of chronostratigraphy has grown in use and has become a standard for long-ranging chronostratigraphic correlations in regions other than Europe. Few attempts have been made, however, to use the European system of classification in the Illinois basin.

At the base of the Middle Carboniferous, the lower and upper stages of Namurian B are named Kinderscoutian and Marsdenian, and the Namurian C is named Yeadonian. The stratotypes of the Namurian have been designated in Great Britain and Ireland (Ramsbottom, 1981). Westphalian A, B, and C also have been named: Langettian, Duckmantian, and Bolsovian, respectively. Their stratotypes have been recognized in Britain (Owens et al., 1985). No formal stratotype has been designated for the Westphalian D. The type area of the overlying Stephanian is in the St. Etienne basin of central France, but the boundary between the Westphalian and Stephanian was defined as the base of the Holz Conglomerate in the Saar–Lorraine region, where some strata are missing. The history of the nomenclature and correlation of the chronostratigraphic units in the Upper Carboniferous in Europe before 1976 was described in detail by Wagner (1976).

In the former Soviet Union, the Carboniferous is divided into three subsystems or divisions: Lower, Middle, and Upper, which have been adopted elsewhere. The Lower Carboniferous corresponds to the Mississippian in the United States and to the

Tournaisian, Visean, and Namurian A of western Europe. The Middle Carboniferous correlates with the Morrowan, Atokan, and Desmoinesian Stages in the United States and the Namurian B and C Stages and Westphalian Series in Europe. The Middle Carboniferous in the former Soviet Union is further divided into the Bashkirian and overlying Moscovian Series and is correlated with strata in western Europe by use of goniatites, fusulinids, brachiopods, and spores. Bouroz et al. (1978) lowered the base of the Bashkirian to make it equivalent to the base of the Pennsylvanian. Fusulinids and plant impressions are most important in zonation of the Upper Carboniferous, which is subdivided into the Kasimovian and Gzhelian Series. The Upper Carboniferous of the former Soviet Union is essentially equivalent to the Missourian and Virgilian Stages (Upper Pennsylvanian) in the United States and the Stephanian in western Europe. Fusulinid studies in Spain (Wagner and Varker, 1971; van Ginkel, 1972) indicate, however, that the base of the Stephanian (including Cantabrian) correlates with the upper part of the underlying Moscovian. Palynological studies (Inosova et al., 1975; Teteryuk, 1974) indicate that the base of the Stephanian, as recognized before 1971, lies within the lower part of the Kasimovian.

Climate, which played a major role in the development of peat-swamp floras in the equatorial coal belt was, for the most part, warm, moist, and tropical. Nevertheless, precipitation varied somewhat through time and areal extent, becoming markedly drier toward the end of the Desmoinesian (Westphalian D). A pronounced change in peat-swamp floras, from lycopod-dominated to fern-dominated, occurred at that time (Phillips et al., 1974, 1985; Phillips and Peppers, 1984; DiMichele et al., 1985; DiMichele and Hook, 1992). The reduction in moisture may have been initiated by several events occurring simultaneously: the northward migration of Pangea, continuation of uplift along the Appalachian orogenic belt, change in the patterns of wind and water currents, development of a monsoonal climate (Parrish, 1982, 1993), widespread glaciation in Gondwana, and major marine regression resulting in extensive emergence of large land masses. A major depletion in atmospheric CO_2 content at about 300 Ma (Budyko, 1982; Berner, 1991) would also have had an adverse effect on the physiology of peat-swamp plants (Peppers, 1986).

Five major groups of plants (lycopods, ferns, seed ferns, sphenopsids, and cordaites) were represented in Pennsylvanian peat swamps or wet lowlands, and each peat-swamp flora was dominated by only a few species. Hydrophytic plants, including lycopod trees of *Lepidophloios, Lepidodendron,* and *Paralycopodites*, were most common in the swamps. Mesophytes, including the lycopod *Sigillaria*, the tree fern *Psaronius*, and the seed fern *Medullosa*, were adapted to slightly drier, better-drained habitats or to substrates that were composed of clastic sediment. Discussions of Pennsylvanian climate and paleoecology, especially within the coal-forming peat mires can be found in Phillips and DiMichele (1981, 1992), Scott and Taylor (1983), Phillips et al. (1985), DiMichele et al. (1985, 1987), Lyons and Alpern (1989), Lyons et al. (1990), Phillips and Cross

(1991), McCabe (1991), DiMichele and Hook (1992), and DiMichele and Phillips (1994).

The cyclicity of Carboniferous sediments was recognized early in North America (Dawson, 1854; Newberry, 1874). The implications of cyclic deposition of Pennsylvanian sediments in Illinois were investigated in the early part of the 1900s (Udden, 1912; Weller, 1930, 1931; Wanless, 1931; Wanless and Weller, 1932) at which time the term cyclothem was introduced in recognition of the repetition of sets of different facies, including coal beds, in the Illinois basin. Attempts were made to correlate coal beds and other rocks in the cyclothems by conducting detailed studies of the composition (including fossils) of each stratum. Some of these thin beds can be traced for many miles because of the continuity of their invertebrate marine fossil assemblages (Wanless, 1957, 1958).

A renewed interest in sequence stratigraphy was initiated by the stratigraphic techniques that made use of seismic reflection data developed by geologists at Exxon Production Research Company in the late 1960s and in the 1970s (Vail et al., 1977; Haq et al., 1987; Posamentier and Vail, 1988). Some of the data used for constructing sea level curves have not been published. Global changes in sea level, combined with coastal onlap and offlap deposition of marine sediments, have been proposed to explain sequence stratigraphy. Because these are global changes, many geologists think that the sequences can be correlated on a global scale. A major question, which has not been adequately addressed, concerns the role sequence stratigraphy plays in nonmarine sequences developed at a distance from shore line. Also, Posamentier and Weimer (1993, p. 737) pointed out that age correlation of coastal sequences may not be precise enough to permit long-distance correlations and that "different age curves from different basins could be stacked to produce meaningless eustatic curves." Intercontinental correlation of cyclostratigraphy has been somewhat successful for Mesozoic and Cenozoic rocks, but correlations of Paleozoic cycles have been less successful because of the extent of erosion and lack of stratigraphic and seismic data. For example, the correlations of major transgressive-regression cycles in the upper Paleozoic proposed by Ross and Ross (1985, 1988) do not agree with the palynological correlations that I propose here.

The pattern of depositional cyclicity has been used in conjunction with detailed paleontological studies to make long-distance correlations in the Midcontinent. Heckel (1986, 1989) primarily used conodonts to correlate cycles between Oklahoma, Kansas, Missouri, Nebraska, and Iowa. Boardman and Heckel (1989) extended the correlation to northern Texas using fusulinids and ammonoid data in addition to conodont data. Connolly and Stanton (1992) made a correlation of Desmoinesian to Missourian transgressive-regressive cycles between Arizona and the Midcontinent using cyclostratigraphy. They concluded that their correlation exceeded the accuracy of biostratigraphy.

Many studies concerning cyclic deposition and cyclostratigraphy have been published in the past 20 years, but a discussion of that research is beyond the scope of this book. The reader may refer to Wilgus et al. (1988) and Posamentier et al. (1993) for discussion of sequence stratigraphy.

The debate continues on whether tectonic activity or the advance and retreat of glaciers in the Southern Hemisphere was the leading cause of large-scale, cyclic deposition during the Pennsylvanian (Weller, 1964; Klein and Willard, 1989; Klein and Kupperman, 1992). Glaciation and climate changes probably played the largest role in the rapid cyclic rise and fall of sea level. Estimates of the extent of sea level change for Pennsylvanian depositional cycles range from about 42 m (138 ft) (Maynard and Leeder, 1992) to 200 m (656 ft) (Ross and Ross, 1987, 1991).

Because this study concerns principally the palynology of coal beds, the precision of the correlations is, at best, within the time elapsed between the deposition of coal-forming peats. This is also true, however, for other kinds of fossils that are restricted to other facies within each Pennsylvanian cyclothem of marine transgression and regression. Stratigraphic sequences in the Carboniferous represent a relatively small percentage of geologic time because of lack of deposition and erosion of older sediment. The largest amount of time in relation to unit thickness in a Pennsylvanian cyclothem is probably recorded by mudstone (underclay or seat rock). These paleosols may have required tens of thousands of years to form while the land surface was emergent following marine regression. Peat swamps or mires that formed on the weathered surface may have been short-lived, or they may have developed over several thousand years and become extensive. Compaction and conversion to coal reduce the thickness of the peat so that coal seams represent a disproportionately large amount of time relative to their thickness. Deposition of thin limestone beds after marine transgression may account for about as much time. Clastic rocks make up by far the largest rock component of cyclothems. Thick sandstone bodies, especially fluvial, may be deposited in a comparatively few years; because of little compaction, they represent a disproportionately small amount of time.

The time span of a minor Pennsylvanian cyclothem in the Midcontinent has been estimated as 44 k.y.; a major cycle may be 235 to 400 k.y. (Heckel, 1986). Estimates are 400 to 450 k.y. for the middle Appalachian region (Chestnut and Cobb, 1989); and 230 to 385 k.y. for the Paradox basin (Goldhammer et al., 1991). Klein (1990) suggested that these time scales are too long because absolute dates obtained from several Carboniferous sanidines extracted from tonsteins indicated that the Pennsylvanian lasted 19 m.y. (Hess and Lippolt, 1986). Harland et al. (1990), however, obtained a length of 32.3 m.y. for the Pennsylvanian, a finding that does not support the shorter absolute dates suggested by Klein (Langenheim, 1991).

ASPECTS OF PALYNOLOGICAL CORRELATIONS

Early subdivisions of Pennsylvanian and Carboniferous strata were made on the basis of studies of plant compressions,

which are still of prime importance. In this century, however, marine megafossils and microfossils (especially fusulinids and conodonts) as well as palynomorphs have become much more widely used in biostratigraphy. Palynological zonation of type or reference sections of Carboniferous strata were usually preceded by a faunal zonation, and zonal boundaries of the two schemes may not coincide. However, the two kinds of zonations complement each other. Major faunal changes usually followed major floral changes. Palynology can be used to a particular advantage where nonmarine rocks occur, where faunas were not preserved, or where evolution within a particular group of animals was slow. Evolution of plants in the Angaran and Gondwanan provincial floras was conservative, but it was sufficiently rapid in the equatorial belt to have produced a large variety of plants and spores, some of which have short stratigraphic ranges.

Owens et al. (1989) cautioned that spores were dispersed from plants growing under a variety of ecological conditions and were deposited in different facies according to the same parameters that determined depositional sites of clastic particles. The distribution of marine faunas, however, is also determined by a wide variety of environmental factors, and a group of fossils may be restricted to a single facies. To narrow the diversity of environments governing the origin and deposition of spores, I am using palynology of coal beds only. Although a variety of environments occur within a peat swamp, spores were widely dispersed by wind and water currents. Shifting environments within the peat swamp would be represented as much as possible by sampling the entire thickness of the coal bed.

Most of the spores macerated from coal were deposited in situ by swamp-dwelling trees, but others were derived from plants growing outside the swamps. Spores and pollen, which were quickly and widely disseminated, were incorporated in the peat along with plant debris and small amounts of mineral matter. This rapid dispersal provides a more nearly synchronous correlation than is possible with most fossil organisms. During the Pennsylvanian, the North American and European plates were in close proximity (Fig. 1) as they coalesced to form part of the great Laurussian land mass (Ziegler et al., 1979; Scotese et al., 1979; Bambach et al., 1980; Scotese, 1986). Thus, the then-shorter distance between the coal basins in North America and Europe reduced the time for migration of floras and dispersal of their spores. Bouroz (1978, p. 25–26) stated, "Spores and pollen are practically independent of local ecological conditions, due to the method of dissemination, which makes for a homogeneity in the quantity and distribution of the various species. Only the time factor alters the palynological composition of different levels, which gives them their high stratigraphical value." Nevertheless, some variations in floral composition occurred from place to place in peat swamps and other environments because of differences in topography, sedimentation, depth of water table, composition of the substrate, and availability of water and nutrients, as well as location relative to water and prevailing wind currents. Correlations made on the basis of palynology of coal beds are more reliable than are those using clastic

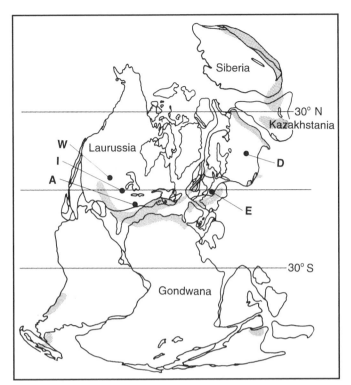

Figure 1. Paleogeographic map of major land masses during the Westphalian. The stratigraphic ranges of spores are illustrated in Plate 2 for the following coal regions: W—Western Interior coal province, I—Illinois basin, A—Appalachian coal region, E—Western Europe, and D—Donets basin. Shading indicates mountain ranges (adapted from Ziegler et al., 1979, 1981; Scotese, 1986).

rocks, because the latter have greater variation in facies and spore assemblages than does coal. Coal beds usually contain an abundance of well-preserved spores, and many of these Carboniferous coals are lithologically continuous over great distances. Redeposited spores are also less likely to occur in coal beds than in clastic sediments.

Palynological correlation of strata between two or more regions is accomplished by comparing palynomorph assemblage zones and the first and last occurrence of spores, as well as spore epiboles and other trends in their relative abundance (Figs. 2 and 3; Plates 2 and 3, in pocket). The first appearance of taxa is usually the most reliable for making correlation between coal basins; for example, the first appearance of *Thymospora* is considered a reliable indicator of the base of the Westphalian D (Plate 2). The last occurrence of some spores is also synchronous, or almost so, as is illustrated by the last occurrence of *Schulzospora rara*, which is used in Europe and North America as an index fossil for the end of the Westphalian A. Changes in the relative abundance of certain spore taxa also have far-reaching significance. These changes are partly dependent on changes in local ecology, tectonics, and depositional patterns, but widespread changes in ecology or adaptive evolution of a plant species may present a competitive advantage over other species for a period of time. For example, the base of the epibole

of *Torispora* is recognized as occurring in middle Westphalian C time, and the base of the epibole of *Thymospora* indicates middle Westphalian D time (Plate 2). The relative abundance of spore populations is more useful in making correlations within a basin because differences in ecology and environments of deposition are more likely to be greater between basins than within a particular basin. The major changes in relative abundance of certain spore taxa in the Illinois Basin (Fig. 2) can be detected throughout the basin.

A comparison of the stratigraphic ranges of species of spores in the lower part of the Pennsylvanian in the Illinois basin, eastern Kentucky, and eastern Tennessee has been useful in making correlations between the Illinois basin and the Appalachian region (Fig. 3). Although stratigraphic ranges of spores are more reliable than relative abundances in making long-distance correlations, the patterns of changes in relative abundance of some of the major spore taxa (Plate 3) are quite similar for the two regions and can be used as additional guides for making correlations. Of particular interest are the patterns in distribution of the most abundant spores: *Radiizonates, Laevigatosporites globosus, Punctatosporites minutus*, and the three most abundant species of *Lycospora*. Distribution patterns of other less common spore taxa may differ from region to region because the ecology in various coal swamps at a given time was not uniform.

Biostratigraphy determined from plant compressions is discussed in this book where appropriate, and stratigraphic ranges of some fusulinids and other microfossils are also mentioned in relation to spore ranges. It is well known that correlations that have been determined from different groups of fossils may not agree. For example, according to evidence from plant compressions (Gillespie and Pfefferkorn, 1979), the top of the New River Formation and Lower Pennsylvanian of the USGS usage is early Westphalian B in age, but according to palynology it is late Westphalian A. Resolution of such inconsistencies is not attempted in this book.

Although the focus of this study is the Illinois basin, published palynological studies of other coal basins were extensively used. Dissertations by students of L. R. Wilson at the University of Oklahoma were also valuable in providing additional data on the palynology of coals in the Midcontinent. I also analyzed numerous samples of coal, especially from near the Atokan-Desmoinesian and Desmoinesian-Missourian boundaries in Oklahoma, Kansas, and Missouri, as an aid in correlation of these boundaries. Most of these samples were collected by me and by P. H. Heckel (University of Iowa) and his colleagues in conjunction with the SEPM Midcontinent Pennsylvanian Stratigraphic Working Group and the Middle Pennsylvanian Working Group of the Subcommission on Carboniferous Stratigraphy, International Union of Geological Sciences.

Major early palynological investigations in the Appalachian coal region include a thesis on Middle Pennsylvanian coals in West Virginia by Schemel (1957) and a thesis on coal beds in Tennessee by Cropp (1958). A summary of the latter was pub-

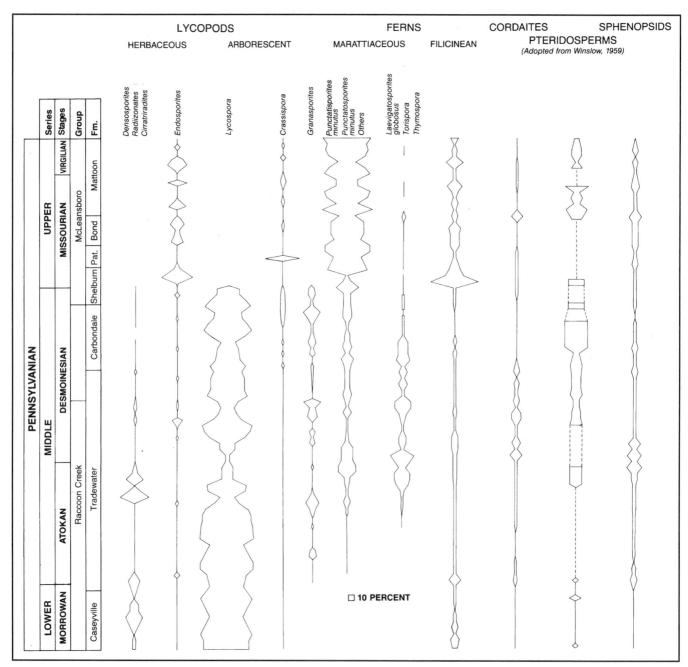

Figure 2. Distribution (in percent) of the most abundant spores in Pennsylvanian coal beds in the Illinois basin. *Punctatisporites minutus*, *Punctatosporites minutus*, and others are thin and smaller than about 25 μm. *Laevigatosporites globosus*, *Torispora*, and *Thymospora* are thicker and larger than about 25 μm. Modified from Phillips et al. (1985).

lished as Cropp (1963). I reexamined the macerations of Tennessee coals reported by Cropp and macerated additional samples that Cropp collected but did not use in his study. Clendening (1974) studied the palynology of coals in the Dunkard Group in West Virginia and concluded from studying the palynology of Upper Pennsylvanian coals in Kansas that the entire group is Pennsylvanian in age. More recently, valuable contributions to the understanding of palynological sequences in eastern Kentucky and West Virginia were made by Eble and Gillespie (1986a, 1986b), Eble and Grady (1990), Eble (1994), Hower et al. (1994), and Kosanke (1973, 1984, 1988a, 1988b, 1988c). I examined coal samples collected recently in the Appalachian coal region, especially from near the Middle-Upper Pennsylvanian boundary in Pennsylvania and Ohio. I also studied numerous coal samples that had been collected from the Appalachian coal region and the Western Interior coal province

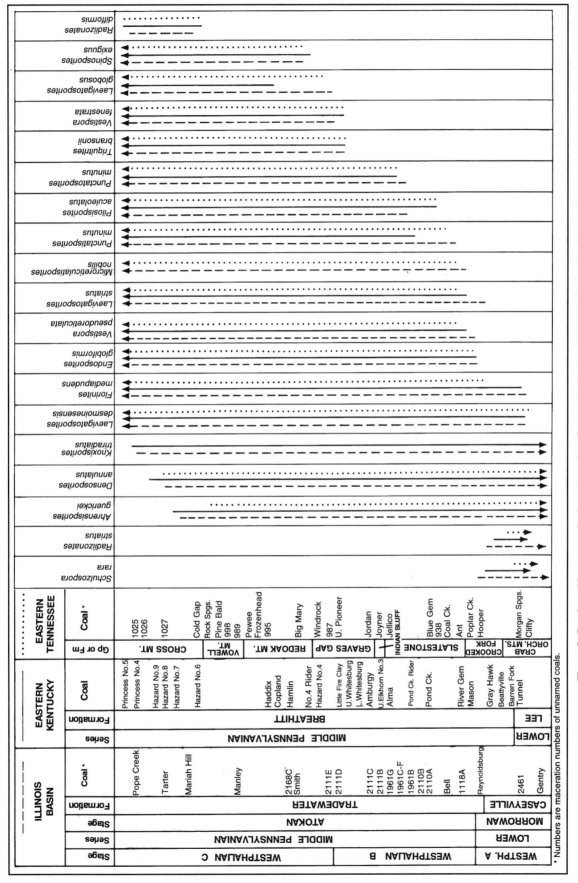

Figure 3. Stratigraphic ranges of selected species of spores in Early and Middle Pennsylvanian coal beds in the Illinois basin, eastern Kentucky, and eastern Tennessee. Unnamed coals are designated by maceration number. Arrows indicate continuous ranges.

and stored (they may no longer be available) in the ISGS collection. Palynological data that have been obtained from numerous studies in the Illinois basin (Schopf, 1938; Kosanke, 1950, 1964; Guennel, 1952, 1958) provided a foundation for the study. More recently, I examined numerous macerations of Lower and Middle Pennsylvanian coal samples collected by mappers working in southern Illinois (Nelson et al., 1991; Jacobson, 1992; Weibel et al., 1993). Palynological analyses of coal samples that I studied specifically for the chronostratigraphic correlations discussed in the text are given in the tables.

I use the stratigraphic nomenclature of the particular state or region being discussed. For example, most states informally designate their coals as coal beds, but they are occasionally formalized. In some states, type sections of coal beds have not been formally designated. The Illinois, Indiana, and Iowa geological surveys treat coals as members for which type sections are designated. This is why "Coal" or "Coal Bed" may be capitalized when referring to a coal in one state, but the same coal is not capitalized when referring to another state. Recently, the ISGS has been reducing the nomenclatural rank of some coals from members to coal beds if they are of local extent. For example, the name of the Reynoldsburg Coal, which can be traced for less than 15 km (9.4 mi), has been changed from Reynoldsburg Coal Member to Reynoldsburg Coal Bed (Trask and Jacobson, 1990).

The text is arranged chronologically, beginning with the oldest major chronostratigraphic boundary discussed. Within each discussion of those boundaries, the region that serves as the type area for the chronostratigraphic divisions is discussed first. My interpretation of the palynological correlation of these boundaries in the other regions is presented initially. It is followed by a review of previous correlations and a discussion of the palynological basis for the correlations. Formal systematics of spore taxa mentioned in the text and listed in the tables is given in Appendix A. The sample locations of coals specifically analyzed for this study are given in Appendix B.

CHRONOSTRATIGRAPHIC CORRELATIONS

Namurian-Westphalian boundary

Western Europe. Butterworth and Millott (1954) were the first to propose a spore zonation for the Carboniferous. Since then, several schemes have been devised, especially in Europe. Most noteworthy is the spore assemblage zonation of coals in the Carboniferous coal basins in Great Britain by Smith and Butterworth (1967). This was followed by the palynological zonation of strata from the Famennian Stage in the Devonian to the Autunian Series in the Permian of western Europe by Clayton et al. (1977). They divided the strata into 25 spore zones from compiled data from numerous studies.

The Heerlen Congress on Carboniferous Stratigraphy and Geology of 1935 defined the base of the Westphalian as the *Gastrioceras subcrenatum* Marine Band. The Subcommission on Carboniferous Stratigraphy has proposed the stratotypes of the Namurian and Westphalian A (Langsettian) to C (Bolsovian)

in outcrops of marine strata in northern Britain and has recorded the palynomorphs contained therein (Ramsbottom, 1981). Smith and Butterworth (1967) did not distinguish any major palynological changes in coal beds at the Namurian-Westphalian boundary, and their *D. annulatus* spore assemblage zone spans the boundary (Plate 1). By studying the spores of lithologies other than coal, Owens et al. (1977) were able to delineate the boundary. Clayton et al. (1977) stated that the Namurian C is characterized by the *Raistrickia fulva-Reticulatisporites reticulatus* (FR) spore assemblage zone (Plate 1), whereas the lower half of the Westphalian A is assigned to the *Cirratriradites saturnii-Triquitrites sinani* (SS) zone. *Punctatisporites pseudopunctatus, Microreticulatisporites punctatus,* and *Cirratriradites rarus* have not been found above the Namurian, whereas *Cingulizonates loricatus* and *Vestispora costata* begin their ranges at the base of the Westphalian. *Cirratriradites saturnii* becomes common in the Westphalian. *Laevigatosporites* occurs sporadically in Namurian B and lower C strata, is not present in uppermost Namurian and in lower Westphalian A, and becomes numerically significant in the upper part of the Westphalian A (Smith and Butterworth, 1967).

Donets basin. Limestone G_1 in the lower part of the Bashkirian Series marks the Namurian-Westphalian boundary in the Donets basin.

Correlation of spore assemblage zones between western Europe and the Donets basin was discussed by Owens et al. (1978) and Owens (1984), but no stratigraphic sections for the Donets basin were included. The Namurian-Westphalian boundary lies between spore zones *Conglobatisporites conglobatus-Mooreisporites trigallerus* (CC-MT) and *Apiculatisporites grumosus-Schulzospora rara* (AG-SR) of Teteryuk (1976). *Knoxisporites seniradiatus, Mooreisporites bellus, M. terjugus, M. trigallerus,* and *Ahrensisporites guerickei* var. *ornatus* disappear in the uppermost Namurian, but the latter extends higher in Europe and North America than in the Donets basin. *Lophotriletes insignitus, Pustulatisporites grumosus,* and *Laevigatosporites minimus* appear at the base of the Westphalian (Teteryuk, 1976).

Appalachian coal region. On the basis of plant compression studies and palynology, I correlate the Namurian-Westphalian boundary in West Virginia with the middle of the New River Formation, just below the Sewell coal bed.

According to Gillespie and Pfefferkorn (1979), the Namurian B and C cannot be differentiated with plant megafossils in the area of the USGS proposed stratotype. Gillespie and Pfefferkorn concluded, however, that the Namurian-Westphalian boundary is near the Beckley coal bed, which is near the middle of the New River Formation. Important plant compressions include *Alethopteris decurrens,* which is restricted to the upper part of the New River Formation, and *Sphenopteris preslesensis,* which is in the lower two-thirds of the formation. Several species of *Neuropteris* disappear in the upper part of the New River. The upper part of the New River Formation is in floral zone 6 of Read and Mamay (1964), and the lower part is in floral zone 5. Wagner (1984) correlated his *Lyginopteris hoening-hausii-Neuralethopteris schlehani* floral zone with parts of Read and Mamay's zones 5 and 6, but he thought that the Pocahontas Formation, which underlies the New River, is Westphalian A in age, as determined by the total range of *Lyginopteris hoening-hausii.* Gillespie and Pfefferkorn (1979) concluded that the Pocahontas Formation is Namurian B in age, and the limited palynological data available support this conclusion.

Most of the spore taxa used to differentiate the Namurian from the Westphalian in Europe have not been found to be very useful in North America. Coal beds in the lower part of the New River Formation and in the Pocahontas Formation in the proposed stratotype do not contain spores, or the spores are opaque or dark as a result of thermal alteration; the number of species that can be identified is small (Kosanke, 1988a). Kosanke observed the earliest occurrence of *Laevigatosporites* in an unnamed coal at the base of the Nuttall Sandstone Member in the upper part of the New River Formation. In 1984 he tentatively identified two specimens as *Laevigatosporites* in the Sewell coal bed, a little above the middle of the New River Formation. Eble et al. (1985) reported that *Laevigatosporites* appears in the No. 3 coal bed of Georgia and the Brookwood coal bed of Alabama, which they correlated with the Sewell coal. According to Clayton et al. (1977), *Laevigatosporites* appears, but is rare, in western Europe at about the Namurian A-B boundary. In Britain (Smith and Butterworth, 1967), its consistent appearance begins in the middle of the Westphalian A, which probably corresponds to its reported first appearance in West Virginia because the middle of the New River Formation is not as old as the Namurian A-B boundary (Mississippian-Pennsylvanian boundary). Thus, the Namurian-Westphalian boundary is at the base of the Sewell coal at the first appearance of *Laevigatosporites* in Alabama and West Virginia and at about the middle of the New River Formation, as also indicated by evidence from plant compressions (Gillespie and Pfefferkorn, 1979).

Illinois basin. The Namurian-Westphalian boundary is correlated in this book with the top of the Battery Rock Sandstone Member in the Caseyville Formation in Illinois, the top of the Kyrock Sandstone Member in western Kentucky, and the lower part of the Mansfield Formation in Indiana (Plate 1).

Although Namurian sedimentary rocks are present in the Illinois Basin, the Namurian-Westphalian boundary has not been well delineated because of poor exposures, stratigraphic discontinuities, variable composition of basal Pennsylvanian sedimentary rocks, and lack of adequate paleontological and palynological information. A major widespread unconformity separates Pennsylvanian strata from Mississippian (Lower Carboniferous) or older strata in the Illinois Basin as well as elsewhere (Saunders and Ramsbottom, 1986). The oldest Pennsylvanian strata were deposited on an erosional surface of deeply incised valleys in the southern and northwestern parts of the basin (Siever, 1951; Bristol and Howard, 1971; Howard, 1979a, 1979b; Leary, 1974, 1975, 1981, 1984, 1985; Droste and Keller, 1989).

Few attempts have been made to correlate the Namurian-Westphalian boundary in the Illinois basin. Shaver (1984) and Shaver et al. (1986) used studies of ostracods and fusulinids to correlate the top of the "Namurian" with the interval between the Mariah Hill and Blue Creek Coal Members of Indiana (middle of the Mansfield Formation) and the lower part of the Abbott Formation (now called lower part of the Tradewater Formation) of Illinois. In the *Correlation Chart for the Midwestern Basin and Arches Region* (COSUNA), Shaver et al. (1985) correlated the boundary with the lower part of the Mansfield and Caseyville Formations. The *Compendium of Indiana Stratigraphy* by Shaver et al. (1986) contains a less accurate correlation of the same boundary than the 1985 correlation chart. Although the 1986 compendium was published after the 1985 chart, parts of the compendium actually might have been written earlier.

White (1896) stated that the "Hindostan Whetstone Beds" (informal name, Shaver et al., 1986) near the base of the Pennsylvanian in southern Indiana are no younger than the Sewell coal bed in West Virginia. As was stated previously in this book, the middle of the New River Formation, which is just below the Sewell coal bed, is correlated with the Namurian-Westphalian boundary. Read and Mamay (1964) concluded that the compression flora in the "Hindostan Whetstone Beds" represents a transition flora between the floras in the New River and underlying Pocahontas Formations in West Virginia; thus, the "Hindostan Whetstone Beds" are probably Namurian C in age. Cross (1992) noted that the roof shale of the French Lick Coal, which underlies the "Whetstone Beds," contains a megaflora equivalent to Pocahontas floras in the Appalachian region, and he also noted that the "Whetstone Beds" contain floras similar to those in the upper Pocahontas to lower New River Formations.

Guennel (1958), who studied the palynology of coals in the lower part of the Pennsylvanian in Indiana, noted that *Laevigatosporites* is not present, but *Schulzospora rara* is abundant in the Pinnick Coal Member, which overlies the "Hindostan Whetstone Beds." Guennel (1958) also reported 15% *Laevigatosporites* and no specimens of *Schulzospora* in the French Lick Coal Member, which underlies the "Hindostan Whetstone Beds." His samples of the Pinnick and French Lick Coals may have been transposed because, in the Illinois basin, *Laevigatosporites* appeared in the upper part of the range zone of *Schulzospora* and did not become abundant until after *Schulzospora* disappeared at the Morrowan-Atokan boundary. Cross (1992) reported the presence of *Schulzospora rara* and *Cingulizonates loricatus* in the French Lick Coal and a much younger spore assemblage in the Pinnick Coal.

The sample of the French Lick Coal that I examined contains *Schulzospora rara* and *Densosporites irregularis* but lacks *Laevigatosporites, Florinites mediapudens,* and *Radiizonates aligerens* (Table 1). *Schulzospora* and *D. irregularis* extend up to the top of the Westphalian A, and *F. mediapudens* appears sporadically in the Namurian A in Europe. It is consistently present at about the middle Westphalian A. The range of *R. aligerens* begins at about middle Westphalian A. An age of

TABLE 1. SPORE ANALYSIS OF FRENCH LICK (MACERATION 151) AND PINNICK (MACERATION 150) COAL MEMBERS IN ORANGE COUNTY, INDIANA*

Spore Taxa[†]	Macerations	
	151 (%)	150 (%)
Deltoidospora sphaerotriangula	X[§]	
Punctatisporites flavus	X	
P. glaber	X	
P. incomptus	X	
P. obesus		X
Calamospora hartungiana	X	
C. liquida	X	
Granulatisporites minutus		0.5
G. pallidus		X
G. tuberculatus	X	
Cyclogranisporites aureus`		0.5
C. minutus		0.5
Verrucosisporites microtuberosus		X
Waltzispora prisca		X
Anapiculatisporites baccatus	1.5	
Spinozonotriletes sp.	X	
Spackmanites habibii		0.5
Convolutispora florida	X	
_C. sp._	0.5	
Ahrensisporites guerickei	X	
Knoxisporites triradiatus	X	
Reticulatisporites polygonalis		X
Reticulitriletes falsus		0.5
Savitrisporites nux		X
Crassispora kosankei	X	
Densosporites annulatus	2.0	X
D. irregularis	X	
D. sphaerotriangularis	1.5	
D. spinifer		X
D. triangularis	X	
D. spp.	1.0	
Lycospora granulata	10.5	56.5
L. micropapillata		1.5
L. noctuina	1.5	0.5
L. orbicula	0.5	
L. pellucida	33.5	2.0
L. pusilla	18.0	16.0
L. rotunda	1.5	
Cingulizonates loricatus	23.5	
Endosporites globiformis		15.0
Schulzospora rara	4.5	
Laevigatosporites desmoinesensis		0.5
L. globosus		X
L. ovalis		1.5
L. vulgaris		0.5
Vestispora costata		X
V. pseudoreticulata		X
Florinites visendus	X	0.5
F. volans		3.0
Potonieisporites elegans	X	
Quasillinites diversiformis		X

*The "Hindostan Whetstone Beds" separate the two coals.
†Formal systematics of taxa given in Appendix A.
§X = present but not in count

Namurian C is indicated for the French Lick Coal and upper Westphalian B for the Pinnick Coal on the basis of palynological studies (Table 1). Thus, a disconformity occurs between the two coals or the "Hindostan Whetstone Beds" represent a long time interval.

Palynology of the thin coal beds in the Upper Mississippian Chesterian in the Illinois basin has been little studied, but the existing data indicate the presence of a gap in the palynological record between the youngest Mississippian and oldest Pennsylvanian strata (Kosanke, 1959a, 1964; Kosanke and Peppers, 1981). Kosanke (1982) reviewed the palynology of strata above and below the Mississippian-Pennsylvanian boundary in the United States. He discussed the stratigraphic ranges of several taxa in several core samples of unnamed coal beds that occur just above the basal Pennsylvanian unconformity in southeastern Illinois. Kosanke (1982) did not indicate an age for these coals in terms of European chronostratigraphy, but he showed that they are older than the Gentry Coal Bed. I have also examined the coals and concur that they are older than the Gentry, probably about the age of the Namurian-Westphalian boundary. The coals contain abundant _Schulzospora_ and _Densosporites irregularis_ but lack _Laevigatosporites_ and _Florinites mediapudens_.

Some paleovalleys cut into pre-Pennsylvanian rocks in northwestern Illinois and eastern Iowa, as an extension of the Illinois basin. These paleovalleys contain coal beds having palynomorph assemblages similar to those in the paleovalleys in southeastern Illinois. Ravn and Fitzgerald (1982) compared these assemblages with those from the Upper Mississippian and Lower Pennsylvanian Springer Group and associated strata in Oklahoma and Texas (Felix and Burbridge, 1967) and from the Pennington Formation and other strata in eastern Kentucky (Ettensohn and Peppers, 1979). The Wildcat Den Coal Member, which is the oldest known Pennsylvanian coal in Iowa, contains _Schulzospora_, _Densosporites irregularis_, and rare specimens of _Laevigatosporites_ and _Florinites mediapudens_. Ravn and Fitzgerald (1982) dated the coal as being at the Namurian-Westphalian boundary, but Ravn (1993, personal communication) now thinks, on the basis of additional studies, that Westphalian A is a more likely age. Therefore, the age of the appearance of _Laevigatosporites_ and _Florinites mediapudens_ in Iowa may not be an anomaly. _Laevigatosporites_ occurs as early as the Namurian in only one of the numerous coal basins in Great Britain studied by Smith and Butterworth (1967).

A paleovalley at the base of the Pennsylvanian in Brown County in eastern Illinois contains well-preserved fossil plant compressions, which were described by Leary and Pfefferkorn (1977). They concluded that the flora is Namurian B in age, and my preliminary investigation of spore assemblages in the same rock (unpublished data) supports that age. Preliminary examination of spore assemblages (Peppers, unpublished data) also indicates a Namurian B age for the thin coal beds and shale just above the Mississippian-Pennsylvanian boundary in a roadcut in Ohio County, western Kentucky (stop 4, Williamson and McGrain, 1979).

TABLE 2. SPORE ANALYSIS OF THE BALDWIN COAL BED IN NORTHERN ARKANSAS*

Spore Taxa[†]	Macerations 2793	Macerations 1443	OWR 1 and 2
Granulatisporites adnatoides		X[§]	
G. sp.	X	X	
Cyclogranisporites sp.	X		
Verrucosisporites microtuberosus			X
Lophotriletes microsaetosus	X		
Apiculatasporites variocorneus	X		
A. sp.	X		
Planisporites granifer			X
Pilosisporites williamsii	X		
Raistrickia saetosa			X
Convolutispora sp.		X	
Ahrensisporites guerickei	X	X	
A. guerickei var. ornatus		X	
Triquitrites sp.			X
Knoxisporites triradiatus	X		
K. spp.			X
Reticulatisporites muricatus			X
R. reticulatus			X
R. sp.			X
Savitrisporites nux		X	X
Crassispora kosankei	X	X	
Densosporites annulatus	X		
D. irregularis	X		
D. sphaerotriangularis		X	
D. triangularis		X	
D. spp.			X
Lycospora micropapillata		X	
L. pellucida		X	
L. spp.			X
Cristatisporites connexus		X	
C. indignabundus	X	X	
Endosporites spp.			X
Schulzospora rara	X	X	
S. spp.			X
Laevigatosporites desmoinesensis	X	X	
L. vulgaris	X		X
Punctatosporites minutus			X
Florinites similis	X		
F. volans	X	X	
F. spp.			X

*As reported in this study (macerations 2793 and 1443) and as reported by Loboziak et al. 1984 (OWR 1 and 2).

†Formal systematics of taxa given in Appendix A.

§X = present but not in count.

almost all of the Desmoinesian in the Western Interior coal basin. The Middle Pennsylvanian includes the upper part of Read and Mamay's (1964) floral zone 6 through zone 10 (*Neuropteris flexuosa* and abundant *Pecopteris*) and Wagner's (1984) floral zone 8 (*Lonchopteris rugosa-Alethopteris urophylla*) through the lower half of floral zone 11 (*Lobatopteris vestita*).

The Lower-Middle Pennsylvanian boundary in southwestern Virginia is at the upper boundary of the Lee Formation but, in the Cumberland Gap area along the Kentucky-Virginia border, it is below the top of the Lee Formation (Englund, 1979). The top of the Lee Formation in Kentucky was tentatively correlated with the top of the Lower Pennsylvanian along and south of Pine Mountain (McDowell et al., 1981). Because of northwestward intertonguing with the overlying Breathitt Formation, however, the Lee extends above the top of the New River Formation in the northwestward direction (Englund, 1961; Rice, 1978). McDowell et al. (1981) indicated that the base of the Corbin Sandstone Member of the Lee Formation is slightly above the tentative position of the Lower-Middle Pennsylvanian boundary in northeastern Kentucky and is considerably above the boundary in east-central Kentucky. The Lee and overlying Breathitt Formations in eastern Kentucky are commonly treated as undifferentiated.

The extinction of *Schulzospora* has become recognized as synchronous, or nearly so, throughout the Euramerican coal belt (Smith and Butterworth, 1967; Loboziak, 1974; Clayton et al., 1977; Owens et al., 1978; Paproth et al., 1983; Peppers, 1984). The stratigraphic range of *Schulzospora* extends above the Lower-Middle Pennsylvanian boundary of the Appalachian region. In West Virginia, the youngest occurrence of *Schulzospora rara* that I observed is in the Lower War Eagle coal bed (Table 3) in the lower part of the Kanawha Formation (Plate 1). The youngest occurrence of some spores Kosanke (1988b) assigned with uncertainty to *Schulzospora* is in an unnamed coal 30.4 m (99.7 ft) above a coal thought to be the Gilbert coal bed. Because the Gilbert coal is a little older than the Lower War Eagle coal, the unnamed coal is probably about the same age as the Lower War Eagle coal. Cross (1947) recognized that a conspicuous change in spore assemblages takes place above the Lower War Eagle coal. Eble and Gillespie (1989) reported that the last occurrence of *Schulzospora* in Virginia is in the Kennedy coal bed at the base of the Middle Pennsylvanian.

In eastern Kentucky, Kosanke (1978, personal communication, *in* Rice and Smith, 1980) observed the youngest occurrence of *Schulzospora rara* in the Gray Hawk coal bed, which is above the Lower-Middle Pennsylvanian boundary in areas along the Pottsville escarpment (Fig. 3). It extends as high as the Splash Dam and Splitseam coal beds (Kosanke, 1982), which are correlative with the Gray Hawk coal, in the lower part of the Breathitt Formation in the easternmost and southeasternmost parts of the Appalachians in Kentucky. I also found *S. rara* in samples of the Gray Hawk and underlying Beattyville and Barren Fork coal beds of eastern Kentucky (Table 3, Fig. 3). A sam-

The age of the oldest Pennsylvanian strata in southern Illinois is not precisely known, but Rexroad and Merrill (1979, 1985) identified conodonts that occur in the basal part of the Pennsylvanian Wayside Member of the Caseyville Formation; they considered these to be morphologically very similar to those in the underlying Mississippian Grove Church Formation. They concluded that deposition may have been continuous locally from the Mississippian through the Pennsylvanian. Jennings (1984) and Jennings and Fraunfelter (1986) noted, however, major differences between the Wayside and Grove Church megafloras and faunas. Weibel and Norby (1990) also found differences between the Mississippian and Pennsylvanian conodonts and assigned the basal Pennsylvanian strata to the early Morrowan: they concluded that an unconformity does separate the Mississippian-Pennsylvanian rocks. My preliminary examination of spores in the Wayside shale indicates that the shale is Namurian in age, probably Namurian B.

Western Interior coal province. The Namurian-Westphalian boundary is here correlated with the basal part of the Bloyd Formation at the Brentwood Limestone of Arkansas at about the middle of the Morrowan (Plate 1).

Ammonoids have been used as the principal basis for chronostratigraphic correlations between the Western Interior coal province and Europe in the lower part of the Pennsylvanian. The work of Saunders et al. (1977) indicates that the Namurian-Westphalian boundary (between ammonoid zones G_1 and G_2) correlates with the lower part of the Brentwood Limestone Member in the lower part of the Bloyd Formation of Arkansas.

Felix and Burbridge (1967) described the palynology of the Springer Formation in the Ouachitas of Oklahoma. The formation extends higher than the Springer Group in the Ardmore basin, and Felix and Burbridge recorded specimens of *Laevigatosporites* in the clastic sequence. They concluded that the Springer Formation is transitional between the Mississippian and Pennsylvanian. Sullivan and Mishell (1971) compared spore assemblages from the Springer and Upper Mississippian strata in southern Oklahoma with European assemblages and concluded that the Springer Formation is no younger than Namurian A. Upper Mississippian and Lower Pennsylvanian spores from clastic sedimentary rocks in Arkansas were also examined by Owens et al. (1984), who agreed with Felix and Burbridge (1967) on the age of the Springer. Sutherland and Manger (1979) concluded that the Springer extends as high as lower Westphalian A.

Miller et al. (1989) reported on the palynology of Upper Mississippian and Lower Pennsylvanian strata at several sites in northeastern Oklahoma. *Laevigatosporites ovalis* and *Florinites mediapudens*, which consistently occur just below the middle of Westphalian A in western Europe, were recorded from middle Morrowan shale in the upper part of the Braggs Member (Sausbee Formation) in northeastern Oklahoma.

The palynological investigation of the Morrowan Hale and Bloyd Formations of Arkansas by Loboziak et al. (1984) recovered only 5 spore taxa from the Prairie Grove Member of the Hale Formation and 12 from the Brentwood Limestone in the lower part of the Bloyd Formation. The Brentwood Limestone i approximately equivalent to the upper part of the Braggs Member. The small number of taxa makes comparison of this spore assemblage and the European palynomorph zones difficult; however, they followed Saunders et al. (1977), who used ammonoids and correlated the Namurian-Westphalian boundary with the lower part of the Brentwood Limestone. *Laevigatosporites* first appears in the Baldwin coal bed (Table 2), which is at the top of the overlying Woolsey Member. According to Sutherland and Manger (1979), the Woolsey Member correlates with the lower part of the Westphalian A. I propose this earliest appearance of *Laevigatosporites* in coal probably corresponds to the beginning of the consistent occurrence of the genus in Europe just below the middle of the Westphalian A. Thus, palynological analysis indicates that the Namurian-Westphalian boundary is a little below the Woolsey Member or at the Brentwood Limestone, a finding that agrees with the correlation as determined from ammonoids.

Lower-Middle Pennsylvanian Series boundary (USGS)

Appalachian coal region. The top of the Lower Pennsylvanian Series in the Appalachian coal region, according to USGS recommendations and usage, is at the top of the New River Formation of West Virginia.

The Appalachian coal region serves as the reference area for defining the boundaries between the Lower, Middle, and Upper Pennsylvanian Series according to USGS terminology (Bradley, 1956). The top of the Lower Pennsylvanian corresponds to the top of the New River Formation of West Virginia where the proposed stratotype of the Pennsylvanian System occurs (Plate 1). The Lower Pennsylvanian Series in West Virginia includes the Pocahontas Formation and overlying New River Formation and is much thicker than the Lower Pennsylvanian in the Illinois basin. Palynology of the lower part of the Lower Pennsylvanian in some parts of the Appalachian coal region is not well known because maturation of the coals has altered the spores beyond recognition.

The Lower Pennsylvanian includes floral zone 4 (*Neuropteris pocahontas* and *Mariopteris eremopteroides*) through the lower part of zone 6 (*Neuropteris tennesseeana* and *Mariopteris pygmaea*) (Read and Mamay, 1964). According to Wagner (1984), the Pocahontas and New River Formations correspond to his *Lyginopteris hoeninghausii-Neuralethopteris schlehani* megafloral zone (7), which he indicated is equivalent to the Westphalian A. As mentioned in the previous section on the Appalachian region, the Pocahontas is, however, Namurian in age.

The Middle Pennsylvanian includes the upper part of the Pottsville Formation (Lesley, 1876) and the Allegheny Formation (Rogers, 1840) as originally defined in the Appalachian region. The Middle Pennsylvanian, as used by the USGS, correlates with the upper part of the Morrowan, the Atokan, and ple from the Hagy coal bed in Pike County, Kentucky, was also studied. The Hagy coal bed is younger than the Gray Hawk coal, but its spores are dark and poorly preserved. Kosanke (1988, personal communication) also didn't observe any specimens of *S. rara* in the Hagy coal. I did not observe *S. rara* in a sample from the Mason coal bed (Bell County, maceration 2783, Table 3), which is a little younger than the Hagy coal. The Gray Hawk, Beattyville, Mason, Splash Dam, and Splitseam coals are above the Rockcastle Sandstone in Kentucky, which is equivalent to the Nuttall Sandstone Member at the top of the New River Formation in West Virginia (Rice et al., 1979a; Rice and Smith, 1980).

The Morgan Springs coal bed in northeastern Tennessee is just below the Rockcastle Sandstone and is the youngest coal examined containing *Schulzospora rara* (Fig. 3, Table 4). No specimens of *S. rara* were observed in the next-younger coal studied, the Hooper coal bed, which is well above the Rockcastle Sandstone (Plate 1) and the top of the Lower Pennsylvanian.

The stratigraphic ranges of *Laevigatosporites, Florinites mediapudens,* and *Densosporites irregularis* are also useful in delineating the Lower-Middle Pennsylvanian boundary. *Laevigatosporites* first appears a little below the Lower-Middle Pennsylvanian boundary in the Appalachian coal region (Fig. 3) and gradually increases in abundance upward into the Middle Pennsylvanian. Kosanke (1984, 1988b) reported that the reliable range of *Laevigatosporites* begins in an unnamed coal a little below the Nuttall Sandstone, but he found two questionable specimens of the genus in the Sewell coal in West Virginia. I found that specimens of *Laevigatosporites* are rare in a sample of the Barren Fork coal (Table 3) from McCreary County; however, Kosanke (1983, personal communication) found that the genus is rather common in his sample of the coal from the Sawyer quadrangle. Kosanke (1982) reported *Laevigatosporites* in the Tunnel coal bed in southeastern Kentucky, a bed that is older than the Barren Fork coal. I also observed *Laevigatosporites* in a sample of the Tunnel coal from Bell County, Kentucky, but not in a sample from the older Stearns coal zone in Wayne County, Kentucky (Table 3). In Ohio, *Laevigatosporites* is well represented in the Quakertown No. 1 coal bed, but it is absent in the underlying Anthony coal bed (Kosanke, 1984). The intervening Huckleberry coal bed has not been studied. Miller (1974) recorded almost 7% *Laevigatosporites* from the Kennedy coal at the base of the Middle Pennsylvanian in southwestern Virginia, and it first appears in the Lee coal bed in the upper part of the Lee Formation (Eble and Gillespie, 1989). The first appearance of *Laevigatosporites* in eastern Tennessee is in the Clifty coal bed (Cropp, 1963).

Florinites mediapudens begins its stratigraphic range soon after *Laevigatosporites* appears and just above the Lower-Middle Pennsylvanian boundary (Plate 2). Kosanke (1988a) first recorded *F. mediapudens* in West Virginia from a coal thought to be the Gilbert. I did not observe the species in a sample of the Lower War Eagle coal (Table 3), which is between the Gilbert

and Eagle coals. *Florinites mediapudens* appears in the Barren Fork coal in Kentucky (Table 3), and Neal (1973) observed it in an unnamed coal a little above the first appearance of *Laevigatosporites* in the lower tongue of the Breathitt (below the base of the Corbin Sandstone Member). *Florinites mediapudens* first appears in the Hooper coal in Tennessee (Fig. 3). It may actually appear somewhat earlier, but no coal samples were available for analysis from the upper part of the Crab Orchard Mountains Group or the lower part of the Crooked Fork Group. In Ohio, *F. mediapudens* is in the seat rock of the Quakertown No. 2 coal but not in the coal proper (Kosanke, 1984).

Densosporites irregularis becomes extinct near the top of the Lower Pennsylvanian, last occurring in the seat rock of a coal at the base of the Nuttall Sandstone in the Pennsylvanian stratotype (Kosanke, 1988b). The coal is at such high rank that the spores are difficult to identify. *Densosporites irregularis* extends up to the Jawbone coal bed in Virginia (Eble and Gillespie, 1989). It last appears in the Beaver Creek and Cumberland Gap coal beds, which are about equivalent to the Tunnel coal, in eastern Kentucky (Kosanke, 1982) and in the Anthony coal of Ohio (Kosanke, 1984).

Eble and Gillespie (1989) reported that *Radiizonates striatus* and *R. aligerens*, which appear before *Laevigatosporites*, occur in the lower to middle part of the Lee Formation in Virginia, but their complete ranges in Virginia are not known. The specimen of *Radiizonates* sp. illustrated by Kosanke (1988b) from just below the Nuttall Sandstone may be *R. aligerens*.

In conclusion, the Early-Middle Pennsylvanian boundary as recognized by the USGS is above the first appearance of *Laevigatosporites* and the disappearance of *Densosporites irregularis*, but below the appearance of *Florinites mediapudens* and the disappearance of *Schulzospora rara*. The Early-Middle Pennsylvanian boundary as used by the USGS is older than the boundary as used in the Western Interior Coal Province.

Illinois basin. Palynological evidence in the Caseyville Formation in Illinois is limited because of abundance of sandstone, but the Lower-Middle Pennsylvanian boundary, as used by the USGS, is correlated here with the base of the Pounds Sandstone Member (Plate 1) above the Gentry Coal Bed. In western Kentucky, I correlate the boundary with the base of the Bee Spring Sandstone Member; in the eastern part of western Kentucky, it is above the Nolin coal bed. No coal equivalent to the Gentry Coal has been identified in Indiana, but the Lower-Middle Pennsylvanian boundary is older than the St. Meinrad Coal Member and younger than the French Lick Coal. The boundary is also correlated with the lower part of the *Schulzospora rara-Laevigatosporites desmoinesensis* (SR) spore assemblage zone (Peppers, 1984).

Among investigations to correlate Pennsylvanian strata in the Appalachian region with the Illinois basin are those of Moore (1944), Cross and Schemel (1952), McKee and Crosby (1975), Rice et al. (1979a), and Phillips (1981), all of whom correlated the top of the New River Formation with the top of the

TABLE 3. SPORE ANALYSIS OF THE STEARNS (MACERATION 2776), BARREN FORK (MACERATION 2742), BEATTYVILLE (MACERATION 2741), GRAY HAWK (MACERATION 2738), AND MASON COAL BEDS (MACERATION 2783) IN KENTUCKY AND THE LOWER WAR EAGLE COAL BED (MACERATION 2763) IN WEST VIRGINIA

Spore Taxa*	Macerations					
	2776 (%)	1742 (%)	1741 (%)	2738 (%)	2763 (%)	2783 (%)
Deltoidospora priddyi	0.5		3.5	X†		
D. sphaerotriangula			X			
D. sp.		X				
Punctatisporites obesus						X
Calamospora breviradiata		X	X			
C. hartungiana		X	0.5	X	X	0.5
C. liquida		X				X
C. spp.	X	0.5				
Granulatisporites adnatoides		0.5	X			
G. granulatus	X			X		
G. micrograniffer			0.5	X	1.0	1.0
G. minutus	1.0	0.5				
G. pallidus			1.5	X	3.0	X
G. verrucosus		0.5				0.5
G. spp.	0.5	0.5	0.5	X	2.0	0.5
Cyclogranisporites aureus			1.5			
C. cf. *aureus*						X
C. minutus						0.5
C. sp.	X					0.5
Verrucosisporites microtuberosus				X	X	
Lophotriletes commissuralis	1.0					
L. pseudaculeatus						0.5
Waltzispora prisca		X	0.5			
Anapiculatisporites minor			0.5			
Pustulatisporites crenatus			X			
Apiculatasporites latigranifer	X					
A. spinososaetosus	X	0.5				
A. variocorneus		X		X		
A. sp.				X		
Raistrickia abdita	X			X		X
R. breveminens					X	X
R. fulva						0.5
Convolutispora mellita	X					
Dictyotriletes bireticulatus			X	X	X	
Ahrensisporites guerickei	0.5		X			
Reinschospora triangularis				X		
Knoxisporites triradiatus		X				
Reticulatisporites reticulatus				X		
Savitrisporites nux	0.5	X	0.5			
Crassispora kosankei		0.1	0.5	X		
Granasporites medius		1.5	4.0	X	2.0	
Densosporites annulatus		2.0	5.0	X	1.0	X
D. duriti					X	
D. irregularis				X		
D. sphaerotriangularis			X	X	8.0	
D. triangularis	X				6.0	
D. sp.			0.5		3.0	

TABLE 3. SPORE ANALYSIS OF THE STEARNS (MACERATION 2776), BARREN FORK
(MACERATION 2742), BEATTYVILLE (MACERATION 2741), GRAY HAWK (MACERATION 2738),
AND MASON COAL BEDS (MACERATION 2783) IN KENTUCKY AND
THE LOWER WAR EAGLE COAL BED (MACERATION 2763) IN WEST VIRGINIA (continued)

Spore Taxa*	Macerations					
	2776 (%)	1742 (%)	1741 (%)	2738 (%)	2763 (%)	2783 (%)
Lycospora brevijuga						0.5
L. granulata	70.0	18.5	28.5	3.5	3.0	9.5
L. micropapillata			0.5		1.0	12.0
L. noctuina				1.0		0.5
L. orbicula		3.5	1.5			0.5
L. pellucida	25.0	63.5	40.5	93.5	59.0	71.0
L. pusilla		4.0	1.5		1.0	
L. rotunda		1.5	X			X
Cristatisporites indignabundus				X		
C. sp.				X		
Cirratriradites maculatus				X		
C. saturni		X				
C. sp.		0.5	0.5			
Cingulizonates loricatus	X		0.5			
Radiizonates striatus			0.5			
Endosporites globiformis					X	X
E. staplinii		X	3.5		1.0	
E. zonalis				X	X	
Schulzospora rara	0.5	X	3.0	X	X	
Alatisporites hexalatus				X		
Laevigatosporites desmoinesensis		X	X			1.5
L. ovalis		1.0	X	2.0	9.0	
L. vulgaris			X			
Vestispora magna						X
Florinites mediapudens		X	X	X		X
F. visendus	0.5					
F. volans	X	X	X			
Wilsonites delicatus		X	X			X
Quasillinites diversiformis			X			
Tinnulisporites cf. *microsaccus*					X	

*Formal systematics of taxa given in Appendix A.
†X = indicates present but not in count.

Caseyville Formation. Rice et al. (1979b) and Phillips et al. (1985) correlated the top of the New River Formation with about the Gentry Coal. McDowell et al. (1981) tentatively correlated the Lower-Middle Pennsylvanian boundary with the Battery Rock coal in western Kentucky, or a little above the middle of the Caseyville Formation. They thought that the coal they called the "Battery Rock coal" of western Kentucky was equivalent to the Gentry Coal (formerly called Battery Rock Coal Member) in Illinois, but it is actually a little younger than the Gentry Coal.

In the Illinois basin, *Schulzospora rara* extends as high as the Reynoldsburg Coal Bed, just above the top of the Caseyville

Formation (Fig. 3); however, it is rare in that coal. Kosanke (1950) noted that *Laevigatosporites* and *Florinites antiquus (F. mediapudens)* are not present in the underlying Gentry Coal but begin their continuous ranges in the Reynoldsburg Coal. A few specimens of *Laevigatosporites* were reported, however, in a coal in the Tar Springs Formation (Chesterian) in the Illinois basin (Kosanke, unpublished data). Examination of a sample (Table 5) of a coal from the Caseyville Formation, mapped as the "Battery Rock coal bed" in the Dekoven quadrangle, Union County, Kentucky (Kehn, 1974), indicates that it is younger than the Gentry Coal (formerly Battery Rock Coal) because it con-

TABLE 4. SPORE ANALYSIS OF SELECTED COAL BEDS IN TENNESSEE

Spore Taxa†	Clifty Coal 1009			Morgan Springs 1012		Hooper 1004	Poplar Creek 1003	Ant 1005	Coal Creek					Coal Creek Rider 988	Blue Gem 990	Jellico 985
Maceration	B* (%)	C (%)	D (%)	A (%)	B (%)	(%)	(%)	(%)	A (%)	B (%)	C 1011 (%)	D (%)	E (%)	(%)	(%)	(%)
Deltoidospora adnata																
D. levis				X					X	X	X§		X			
D. priddyi									3.0	1.5	0.5	1.5	X			
D. sphaerotriangula									X	0.5			0.5			
D. subadnatoides									3.0	0.5	0.5		X			
D. spp.				1.0									0.5			
Punctatisporites glaber																
P. minutus						X				0.5	X					
P. obesus					0.5					0.5			X	X	X	0.5
P. sp.							1.0									
Calamospora breviradiata			1.0					0.5				0.5	1.0	X	1.0	1.5
C. hartungiana			X		1.0		X	1.0				X	X	X	0.5	
C. liquida			X													
C. mutabilis													X			
C. pedata													X			
C. spp.							1.0			0.5	0.5		X			
Granulatisporites adnatoides				X									X	X		1.0
G. granularis				X			X			1.5	1.5	0.5	0.5	0.5		
G. granulatus	1.5												1.0			0.5
G. minutus		2.0	1.0	5.0	X			1.0	X	2.5		X	1.0		X	
G. pallidus				5.0	8.5			1.5	1.0						X	1.5
G. piroformis			1.0					1.0				0.5				
G. verrucosus		1.0								1.0	X	0.5			X	
G. spp.	4.0			1.0	1.5	1.0	3.0	0.5	1.0	1.0					0.5	
Cyclogranisporites aureus						X		X								
C. leopoldi													X			
C. minutus								X								
C. spp.				X			2.0									
Sinuspores sinuatus				X				X	X			X	X			0.5
Verrucosisporites microtuberosus								X		2.0	X	0.5	X			
V. sifati												X	1.0			
V. verrucosus								X		1.0		X	X			
V. cf. verrucosus														X	X	
V. spp.				X												

TABLE 4. SPORE ANALYSIS OF SELECTED COAL BEDS IN TENNESSEE (continued - page 2)

Spore Taxa†	Clifty Coal B* (%)	Clifty Coal 1009 C (%)	Clifty Coal D (%)	Morgan Springs 1012 A (%)	Morgan Springs 1012 B (%)	Hooper 1004 (%)	Poplar Creek 1003 (%)	Ant 1005 (%)	Coal Creek 1011 A (%)	Coal Creek 1011 B (%)	Coal Creek 1011 C (%)	Coal Creek 1011 D (%)	Coal Creek 1011 E (%)	Coal Creek Rider 988 (%)	Blue Gem 990 (%)	Jellico 985 (%)
Lophotriletes commissuralis						1.0	X	X						X		
L. gibbosus																X
L. microsaetosus			X	0.5					2.0	0.5		2.0	1.0	0.5		
L. mosaicus											0.5	0.5	0.5			
L spp.										1.0		2.0	1.5			
Waltzispora prisca					X											
Anapiculatisporites baccatus				2.5												
A. spinosus		1.0						0.5								
Pustulatisporites sp.										0.5						1.0
Apiculatasporites latigranifer	0.5															
A. setulosus				X	X				X						X	
A. spinososaetosus				0.5					X							
A. variocorneus											X		X	X		
A. spp.				X	0.5			X	1.0		X				0.5	
Planisporites granifer				X	X					0.5						
Pilosisporites aculeolatus														X	X	
P. williamsii										X						
P. spp.		1.0											0.5			
Raistrickia abdita																
R. aculeolata												0.5		X		
R. breveminens											X			0.5	X	X
R. crocea					X		X	X	X			0.5	X		X	
Spackmanites habibii																
Convolutispora florida											X		X		X	X
C. mellita					X								X			
Microreticulatisporites harrisonii												X	X			
M. nobilis												X				
Dictyotriletes bireticulatus									1.0		X		X	X	0.5	X

TABLE 4. SPORE ANALYSIS OF SELECTED COAL BEDS IN TENNESSEE (continued - page 3)

Spore Taxon	Clifty Coal B* (%)	Clifty Coal C 1009 (%)	Clifty Coal D (%)	Morgan Springs A 1012 (%)	Morgan Springs B (%)	Hooper 1004 (%)	Poplar Creek 1003 (%)	Ant 1005 (%)	Coal Creek A (%)	Coal Creek B (%)	Coal Creek C 1011 (%)	Coal Creek D (%)	Coal Creek E (%)	Coal Creek Rider 988 (%)	Blue Gem 990 (%)	Jellico 985 (%)
Camptotriletes bucculentus					X										X	
Ahrensisporites guerickei		X	X	X						X	X	X	X		X	X
A. guerickei var. *ornatus*							X	0.5	6.0							
Triquitrites additus											X					
T. bransonii										X	X		X			
T. tribullatus								X		X	X	X	X			X
T. spp.			X							1.0						
Zosterosporites triangularis									X			X				
Mooreisporites fustus									X	X			X			
Reinschospora magnifica	0.5	1.0								X					X	
R. triangularis									X				0.5			
Knoxisporites stephanephorus										0.5						
K. triradiatus				X					X		X	X				
Reticulatisporites muricatus													X	X	X	
R. polygonalis										X	X	X	0.5	X	X	
R. reticulatus											0.5			1.0		X
R. reticulocingulum										X	0.5	X				
Reticulitriletes falsus				X			X			X	0.5	X	X			
Savitrisporites asperatus									X						X	
S. nux			X												X	X
Grumosisporites varioreticulatus											X	X				
Crassispora kosankei	2.5				X			0.5		0.5		0.5			X	
Granasporites medius				1.0				0.5		0.5	1.0	0.5	0.5	1.0	0.5	2.0
Simozonotriletes intortus									X				X			

TABLE 4. SPORE ANALYSIS OF SELECTED COAL BEDS IN TENNESSEE (continued - page 4)

Spore Taxa†	Clifty Coal 1009 B* (%)	Clifty Coal 1009 C (%)	Clifty Coal 1009 D (%)	Morgan Springs 1012 A (%)	Morgan Springs 1012 B (%)	Hooper 1004 (%)	Poplar Creek 1003 (%)	Ant 1005 (%)	Coal Creek 1011 A (%)	Coal Creek 1011 B (%)	Coal Creek 1011 C (%)	Coal Creek 1011 D (%)	Coal Creek 1011 E (%)	Coal Creek Rider 988 (%)	Blue Gem 990 (%)	Jellico 985 (%)
Densosporites annulatus	0.5				0.5	13.0	8.0	X	1.0	2.5	2.5		1.5	X	X	X
D. sphaerotriangularis		X	X	0.5	0.5	27.0			1.0	0.5	0.5	0.5	X	1.0		4.5
D. triangularis		X			X				40.0	4.0	1.0		3.5			
D. variabilis				X	X											
D. spp.		1.0		0.5	2.5					1.0			0.5			
Lycospora granulata	29.5	15.0	11.0	5.0	3.0	15.0	17.0	12.5	1.0	6.0	4.5	8.5	2.5	1.5	11.0	8.5
L. micropapillata	2.5			1.0	0.5		1.0	1.5			1.0		0.5		0.5	
L. noctuina	X			0.5	X		X									
L. orbicula		3.0	3.0													2.0
L. pellucida	53.5	73.0	81.0	72.0	80.0	33.0	51.0	72.5	4.0	12.5	47.5	52.5	55.0	91.5	76.5	60.5
L. pusilla			2.0			5.0	5.0	0.5		3.5	1.5	2.0			2.5	4.0
L. rotunda	4.0	1.0					2.0	1.5		0.5		0.5	0.5		0.5	
L. subjuga			X									X				
Cristatisporites connexus	0.5			0.5						1.0	0.5		0.5			
C. indignabundus					X		X			X	1.0	0.5	4.5	0.5		
Paleospora fragila										X	X					
Cirratriradites saturnii													X	1.0		
Cingulizonates loricatus						3.0										
Radiizonates striatus																
Endosporites globiformis								X	X		0.5	1.0		0.5	0.5	1.5
E. staplinii				X									X			0.5
E. zonalis				X		X			X	X	X		0.5		X	X
Schulzospora rara	X	1.0	X	X												
Alatisporites hoffmeisterii										X	X					
A. sp.																
Proprisporites laevigatus			X													
Hymenospora multirugosa							X									
Laevigatosporites desmoinesensis				3.5	0.5	1.0	1.0	1.5	5.0	19.0	11.5	1.5	1.0		1.5	3.0
L. medius										0.5		1.0	1.0		0.5	
L. ovalis		X				1.0	7.0	2.5	20.0	20.5	20.5	13.0	14.5	1.0	3.0	5.5
L. striatus											X		X			

TABLE 4. SPORE ANALYSIS OF SELECTED COAL BEDS IN TENNESSEE (continued - page 5)

| Maceration | Clifty Coal 1009 | | | Morgan Springs 1012 | | Hooper 1004 | Poplar Creek 1003 | Ant 1005 | Coal Creek 1011 | | | | | Coa Creekl Rider 988 | Blue Gem 990 | Jellico 985 |
Spore Taxa†	B* (%)	C (%)	D (%)	A (%)	B (%)	(%)	(%)	(%)	A (%)	B (%)	C (%)	D (%)	E (%)	(%)	(%)	(%)
L. vulgaris									X	X		X	0.5		X	
Punctatosporites minutus					0.5											0.5
Vestispora costata									X			X	X			
V. pseudoreticulata									X					X		
V. spp.															X	
Florinites																
mediapudens	0.5						1.0		5.0	7.5		5.0	3.0			0.5
F. visendus		X													X	X
F. volans			X	X						2.5		X				
F. spp.									4.0		2.0	3.0		0.5		
Wilsonites circularis												0.5				
W. vesicatus									1.0			0.5			X	
Potonieisporites elegans									X	0.5						
Pityosporites westphalensis											X					
Peppersites ellipticus						X					X					X
Trihyphaecites triangulatus																X
Botryococcus sp.											X					

*Spores in maceration 1009A poorly preserved.
†Formal systematics of taxa given in Appendix A.
§X = present but not in count.

TABLE 4. SPORE ANALYSIS OF SELECTED COAL BEDS IN TENNESSEE (continued - page 6)

Maceration	Joyner 1017			Jordon 981	Upper Pioneer 1015	Unnamed Coal 987	Windrock 1006					Big Mary 2665				Unnamed Coal 995
Spore Taxa*	A (%)	B (%)	C (%)	(%)	(%)	(%)	A (%)	B (%)	C (%)	D (%)	E (%)	A (%)	B (%)	C (%)	D (%)	(%)
Deltoidospora levis											X†					
D. priddyi			0.5	0.5		X		1.0	2.0		4.0	X	X	1.0	X	0.5
D. sphaerotriangula				X										X		0.5
D. subadnatoides				X				X		1.0						
D. sp.								X								
Punctatisporites cf. *edgarensis*											1.0					
P. flavus				X												
P. glaber				X										X		
P. minutus		X		1.0						1.0	1.0			1.0	1.0	1.5
P. obesus							X									0.5
Calamospora breviradiata		X				0.5									0.5	0.5
C. hartungiana		X	X	X	X	X	X		1.0	X		X		X		X
C. liquida			X		X		X							X		
C. mutabilis			X			X	X			1.0						
Granulatisporites adnatoides			X			0.5		1.0		1.0			0.5		1.0	0.5
G. granularis	0.5		X						X	1.0			X			
G. granulatus				X				1.0	1.0	1.0	1.0		0.5		2.5	X
G. minutus			X	X	X		1.0	X		7.0	2.0				1.5	
G. pallidus			X	X	1.5	0.5					2.0				X	0.5
G. piroformis												X				1.5
G. verrucosus		X			2.5										X	
G. spp.				1.0	1.0				1.0	1.0				2.0	2.5	
Cyclogranisporites aureus		X					1.0			X		X	X			1.5
C. minutus			X			X										1.0
C. spp.			1.0													
Sinuspores sinuatus							X	X			X			X		
Converrucosisporites sp.								X				X				0.5
Verrucosisporites microtuberosus			X	0.5	1.0			1.0					X			
V. sifati			X	X			X									
V. verrucosus			X	X				1.0			1.0			X	X	0.5
V. cf. *verrucosus*																X
V. spp.								1.0	1.0							

TABLE 4. SPORE ANALYSIS OF SELECTED COAL BEDS IN TENNESSEE (continued - page 7)

All values are in percent (%).

Maceration →	Joyner 1017			Jordon 981	Upper Pioneer 1015	Unnamed Coal 987	Windrock 1006					Big Mary 2665				Unnamed Coal 995
Spore Taxa*	A	B	C				A	B	C	D	E	A	B	C	D	
Lophotriletes commissuralis			X				1.0							1.0	1.5	
L. granoornatus		X											X			
L. microsaetosus		X	0.5	1.0	X	0.5	1.0	3.0	3.0		5.0	1.0	0.5	1.0	1.5	1.0
L. mosaicus					0.5											
L. pseudaculeatus				X								X				
L. rarispinosus											X					
L. spp.						1.0	X				1.0					0.5
Anapiculatisporites spinosus							1.0	1.0		1.0			X		0.5	
Pustulatisporites crenatus							X									
P. sp.												X				
Apiculatasporites spinososaetosus			X											X		
Planisporites granifer							X	X								
Pilosisporites aculeolatus			X	3.5		X		1.0	X	2.0	1.0					
P. williamsii				0.5			1.0	2.0	X		X					
P. spp.				X		0.5					1.0		X			0.5
Raistrickia abdita				X											X	
R. breveminens	0.5							X						1.0		X
R. crocea																X
Spackmanites habibii																
Convolutispora florida								X			X				X	
C. sp.			X													
Microreticulatisporites harrisonii		X														
M. sulcatus								X		X						0.5
Dictyotriletes bireticulatus				4.0			X		X	X						
D. sp.								X	X							
Camptotriletes bucculentus															X	
Ahrensisporites guerickei				X				X					X	X		
A. guerickei var. ornatus					X		X			X					X	X

TABLE 4. SPORE ANALYSIS OF SELECTED COAL BEDS IN TENNESSEE (continued - page 8)

Spore Taxa*	Joyner 1017 A (%)	Joyner 1017 B (%)	Joyner 1017 C (%)	Jordon Pioneer 981 (%)	Upper Coal 1015 (%)	Unnamed 987 (%)	Windrock 1006 A (%)	Windrock 1006 B (%)	Windrock 1006 C (%)	Windrock 1006 D (%)	Windrock 1006 E (%)	Big Mary 2665 A (%)	Big Mary 2665 B (%)	Big Mary 2665 C (%)	Big Mary 2665 D (%)	Unnamed Coal 995 (%)
Triquitrites additus				0.5							1.0					
T. bransonii															1.5	
T. crassus					X		X		X					X	X	
T. sculptilis																0.5
T. tribullatus				X			1.0								X	
T. spp.															1.0	
Reinschospora magnifica							X									
Knoxisporites triradiatus		X					X							2.0	X	
Reticulatisporites mediareticulatus					X											
R. muricatus									X						X	
Reticulitrietes falsus					0.5				X							X
Savitrisporites asperatus																
S. nux		X		X	X					X		X			X	
Grumosisporites varioreticulatus																
Crassispora kosankei								X					X	2.0		X
Granasporites medius	0.5		3.5			X	8.0		1.0			1.0		1.0	X	
Simozonotriletes intortus							X									
Densosporites annulatus				X	1.0	X	1.0		X			13.0		7.0		
D. glandulosus				10.0											X	
D. sphaerotriangularis			X	13.0								4.0	X			
D. triangularis														1.0		
D. spp.							2.0							1.0		
Lycospora brevijuga			0.5													0.5
L. granulata	61.0	56.5	37.0	4.5	25.5	20.0	28.0	43.0	25.0	21.0	20.0	34.0	42.5	36.0	22.5	51.5
L. micropapillata	2.0	3.5		1.0		46.0	14.0	7.0	5.0	2.0	4.0	31.0	44.0	13.0	33.5	3.0

TABLE 4. SPORE ANALYSIS OF SELECTED COAL BEDS IN TENNESSEE (continued - page 9)

Spore Taxa*	Joyner 1017 A (%)	B (%)	C (%)	Jordon 981 (%)	Upper Pioneer 1015 (%)	Unnamed Coal 987 (%)	Windrock 1006 A (%)	B (%)	C (%)	D (%)	E (%)	Big Mary 2665 A (%)	B (%)	C (%)	D (%)	Unnamed Coal 995 (%)
L. orbicula	10.0	7.5	5.0	23.5	1.0	1.5										0.5
L. pellucida	21.0	30.0	37.5		2.0	26.5	21.0	22.0	47.0	32.0	11.0	10.0	10.0	6.0	11.5	11.0
L. pusilla	2.0	0.5	0.5	1.5		0.5		7.0		4.0		1.0	1.0	3.0	2.0	1.0
L. rotunda	1.0			X						1.0					1.0	
L. spp.																0.5
Cristatisporites connexus							X									
C. indignabundus			X	0.5			X								X	X
Paleospora fragila				X												
Cirratriradites maculatus							X							X		
C. saturnii			X								X				X	1.0
C. sp.												1.0				
Endosporites globiformis	0.5		1.0	1.0		0.5	1.0				2.0		X		0.5	3.0
E. staplinii			X	0.5			X	X		X						
E. zonalis				X			1.0	2.0	X							
Alatisporites pustulatus										X	X					
Laevigatosporites desmoinesensis	0.5	1.0		5.5	5.5		8.0	1.0		2.0	X	X	X		1.0	0.5
L. globosus								X	X	1.0	X		X	X	0.5	1.5
L. medius				1.0	1.0	0.5		2.0	2.0	1.0	1.0					
L. ovalis		0.5	7.0	20.5	50.5	1.0	7.0	2.0	5.0	8.0	20.0	3.0	1.0	11.0	8.0	7.5
L. striatus				X	0.5		1.0			X						X
L. vulgaris				1.0	X		1.0	X		X						
Renisporites confossus											X					
Latosporites minutus									2.0	4.0	2.0					1.0
Punctatosporites minutus				1.0	0.5				4.0	3.0	10.0	X	X	3.0	3.5	2.5
Spinosporites exiguus								2.0						1.0		
Vestispora fenestrata					X	X				X						
V. magna	X															
V. pseudoreticulata			X	X		X	X						X		X	

TABLE 4. SPORE ANALYSIS OF SELECTED COAL BEDS IN TENNESSEE (continued - page 10)

	Joyner 1017			Jordon 981	Upper Pioneer 1015	Unnamed Coal 987	Windrock 1006					Big Mary 2665				Unnamed Coal 995
Maceration	A	B	C				A	B	C	D	E	A	B	C	D	
Spore Taxa*	(%)	(%)	(%)	(%)	(%)	(%)	(%)	(%)	(%)	(%)	(%)	(%)	(%)	(%)	(%)	(%)
Florinites																
mediapudens																
F. similis	1.0		3.0		X					3.0		1.0	X	3.0	1.0	0.5
F. visendus		X	2.5	3.0	X		X			X	X	X		1.0	X	
F. volans							1.0	1.0	X	1.0	X			X		
F. spp.					6.0						9.0					1.5
Wilsonites																
delicatus										1.0						
W. vesicatus			0.5	X										1.0		0.5
Potonieisporites																
solidus							X									
Pityosporites																
westphalensis							X									
Peppersites																
ellipticus				X												
Botryococcus sp.						X	X									

*Formal systematics of taxa given in Appendix A.
†X = present but not in count.

TABLE 4. SPORE ANALYSIS OF SELECTED COAL BEDS IN TENNESSEE (continued - page 11)

Spore Taxa*	Frozenhead 997 A (%)	Frozenhead 997 B (%)	Pewee 996 A (%)	Pewee 996 B (%)	Pewee Rider 989 (%)	Unnamed Coal 998 A (%)	Unnamed Coal 998 B (%)	Pine Bald 1030 A (%)	Pine Bald 1030 B (%)	Rock Springs 1029 A (%)	Rock Springs 1029 B (%)	Rock Springs 1029 C (%)	Cold Gap 1028 (%)	Unnamed Coal 1027 (%)	Unnamed Coal 1026 (%)	Unnamed Coal 1025 A (%)	Unnamed Coal 1025 B (%)
Deltoidospora adnata																	
D. priddyi	3.0	0.5	0.5	1.0	0.5	1.5	X	X	X† 1.0	2.0	3.5	1.0	2.0	1.5		0.5	2.5
D. sphaero-triangula	0.5								X	X		0.5			X		
D. subadna-toides		0.5			0.5					1.0				0.5		0.5	0.5
D. sp.																	
Punctatisporites glaber				0.5													
P. minutus		0.5	X	0.5	1.0			2.0	1.0	4.0	10.0		1.5	0.5	0.5		
P. obesus	0.5		0.5			X			X	X		X	X	0.5		X	
P. spp.			0.5														0.5
Calamospora breviradiata	0.5		0.5	0.5	1.5			1.0			2.0	2.5	2.0	1.0			
C. hartungiana			X					0.5	1.0						1.0		0.5
C. liquida			X					X		X							
C. mutabilis		0.5										X				X	X
C. straminea							X					X		X	X		
C. spp.		0.5	0.5							0.5				0.5		0.5	0.5
Granulatisporites adnatoides	0.5							X			0.5	1.5	2.5			0.5	2.0
G. granularis	0.5	X						1.0								X	0.5
G. granulatus	1.5																
G. minutus			X		1.5					0.5	2.0	2.0	0.5			X	1.5
G. pallidus			0.5	2.5	0.5		1.0	0.5	2.0	1.0	0.5	2.0	1.0		1.0		0.5
G. piroformis	0.5												0.5				
G. verrucosus			0.5						1.0				0.5				
G. spp.	1.0	0.5				0.5		0.5		0.5	1.0		1.5		0.5	0.5	
Cyclogranisporites aureus	0.5		0.5	0.5			X	X	X				0.5		0.5	0.5	0.5
C. leopoldi								0.5									0.5
C. minutus					X									0.5		X	X
C. obliquus														0.5			
C. orbicularis	1.0													0.5			
Converruco-sisporites sp.			1.0				0.5		1.0								

TABLE 4. SPORE ANALYSIS OF SELECTED COAL BEDS IN TENNESSEE (continued - page 12)

Maceration → Spore Taxa*	Frozenhead 997 A (%)	997 B (%)	Pewee 996 A (%)	996 B (%)	Pewee Rider 989 (%)	Unnamed Coal 998 A (%)	998 B (%)	Pine Bald 1030 A (%)	1030 B (%)	Rock Springs 1029 A (%)	1029 B (%)	1029 C (%)	Cold Gap 1028 (%)	Unnamed Coal 1027 (%)	Unnamed Coal 1026 (%)	Unnamed Coal 1025 A (%)	1025 B (%)
Verrucosisporites microtuberosus	1.5		1.0	0.5	0.5	X			1.0		1.0				3.5	1.5	0.5
V. sifati	X	X	0.5	0.5	0.5	2.0			X						0.5	1.5	
V. verrucosus	0.5		0.5	1.5	0.5	X	X	X				X		X			0.5
V. cf. verrucosus	0.5				0.5											X	X
V. spp.	0.5					2.0			1.0			0.5					
Lophotriletes commissuralis		0.5			0.5			0.5			1.0	1.5			0.5	0.5	
L. copiosus													X				
L. gibbosus										X							
L. granoornatus			X	X													
L. microsaetosus	X	X		1.0			X	X			X	X			X	X	0.5
L. mosaicus				0.5									X				
L. pseudaculeatus												X					
L. rarispinosus	X		X									X				X	
L. sp.													0.5		0.5		
Anapiculatisporites spinosus																	
Pustulatisporites crenatus							X					0.5					
Apiculatasporites latigranifer															X		
A. setulosus				0.5	X	X								0.5			
A. spinososaetosus		X	0.5											X			
A. variocorneus	X																
A. sp.							X				1.0		X				
Planisporites granifer	X		X			X											
Pilosisporites aculeolatus	6.0		0.5	4.5				X						1.0			
P. williamsii	2.0	0.5	0.5	0.5					1.0								
P. spp.	0.5										0.5						
Procoronsporites sp.	X									X							
Raistrickia abdita																	X
R. aculeata												0.5					

TABLE 4. SPORE ANALYSIS OF SELECTED COAL BEDS IN TENNESSEE (continued - page 13)

Spore Taxa*	Frozenhead 997 A (%)	Frozenhead 997 B (%)	Pewee 996 A (%)	Pewee 996 B (%)	Pewee Rider 989 (%)	Unnamed Coal 998 A (%)	Unnamed Coal 998 B (%)	Pine Bald 1030 A (%)	Pine Bald 1030 B (%)	Rock Springs 1029 A (%)	Rock Springs 1029 B (%)	Rock Springs 1029 C (%)	Cold Gap 1028 (%)	Unnamed Coal 1027 (%)	Unnamed Coal 1026 (%)	Unnamed Coal 1025 A (%)	Unnamed Coal 1025 B (%)
R. breveminens														X			
R. crocea					X							0.5	X	0.5			X
Spackmanites habibii	X		X	X	X			X	X								
Microreticulatisporites harrisonii	0.5		1.0			0.5		X									
M. nobilis		X				0.5							1.0		0.5	X	
M. sulcatus						X						X			X	X	X
Dictyotriletes bireticulatus	0.5	X			X		X		X								
Camptotriletes bucculentus													0.5				
Ahrenisporites guerickei					0.5		X			X							
A. guerickei var. ornatus					X				X								
Triquitrites additus	0.5					0.5	0.5										
T. bransonii	0.5	X		1.5	4.0	0.5	0.5		1.0				0.5				
T. exiguus					X												
T. protensus					X										X		
T. sculptilis					X	1.0	1.5		X	1.0		0.5	5.5	X	2.0	X	1.0
T. subspinosus	0.5																
T. tribullatus	1.5																
T. spp.				0.5	0.5	X	0.5										
Reinschospora magnifica																	
Reticulatisporites muricatus								X									
R. polygonalis				X		X			X	X					X	X	
R. reticulatus	0.5		0.5			X	X		X	X		0.5		X	0.5		
R. reticulocingulum					X			0.5							0.5	X	
Savitrisporites nux	X		X							X		X					
Crassispora kosankei			1.5							X						0.5	

TABLE 4. SPORE ANALYSIS OF SELECTED COAL BEDS IN TENNESSEE (continued - page 14)

Maceration	Frozenhead 997		Pewee 996		Pewee Rider 989	Unnamed Coal 998		Pine Bald 1030		Rock Springs 1029			Cold Gap 1028	Unnamed Coal 1027	Unnamed Coal 1026	Unnamed Coal 1025	
Spore Taxa*	A (%)	B (%)	A (%)	B (%)	(%)	A (%)	B (%)	A (%)	B (%)	A (%)	B (%)	C (%)	(%)	(%)	(%)	A (%)	B (%)
Granasporites medius	2.5	2.5	1.0	0.5	1.0	1.0					1.0	0.5				0.5	
Densosporites annulatus	1.0	17.5	8.5	0.5	1.0		X	X	1.0	X		X		1.5		X	
D. sphaero-triangularis				0.5													
D. triangularis			0.5	X													
Lycospora brevijuga		0.5												1.5			
L. granulata	6.5	30.5	15.5	12.5	25.0	41.0	58.0	49.0	13.0	12.5	26.5	46.5	36.5	33.5	4.0	13.0	31.0
L. micropapillata		1.0	3.0	0.5	2.0	4.0	5.0	0.5		10.0	6.0	7.0	2.5	1.0		1.0	2.5
L. noctuina					0.5					0.5		0.5					
L. orbicula		2.0				4.5	4.0					0.5		0.5			
L. pellucida	1.5	11.5	23.0	7.5	3.5	10.0	23.0	12.0	3.0	35.0	16.0	19.0	16.5	14.0	X	2.0	35.5
L. pusilla			1.0	1.0		0.5	0.5	6.5	2.0	2.0		0.5	0.5	0.5		6.5	2.5
Cristatisporites indignabundus		2.0		1.0													
Paleospora fragila	X		X													X	
Cirratriradites annuliformis								X									
C. maculatus			X			X	X							X			
C. saturnii	1.5		X	0.5		0.5	0.5	0.5		X	X	1.0		X		X	
Radiizonates difformis													X		X	12.5	
Endosporites globiformis		1.5	0.5	2.0	1.0	0.5		5.5	8.0					2.0			
E. plicatus								X								X	
E. zonalis					0.5	0.5											
Alatisporites hoffmeisterii																	
A. pustulatus						X	X			X						X	
A. trialatus					X												X
Laevigatosporites desmoinesensis	4.5	2.0	3.5	1.5	1.5	4.0		0.5	5.0			0.5		1.5	3.5		
L. globosus			2.0	3.0			X		X		1.0			4.5	5.5	44.5	1.5
L. medius	0.5		1.0	0.5	1.0			X	X					X	0.5		

TABLE 4. SPORE ANALYSIS OF SELECTED COAL BEDS IN TENNESSEE (continued - page 15)

Spore Taxa*	Frozenhead 997 A (%)	Frozenhead 997 B (%)	Pewee 996 A (%)	Pewee 996 B (%)	Pewee Rider 989 (%)	Unnamed Coal 998 A (%)	Unnamed Coal 998 B (%)	Pine Bald 1030 A (%)	Pine Bald 1030 B (%)	Rock Springs 1029 A (%)	Rock Springs 1029 B (%)	Rock Springs 1029 C (%)	Cold Gap 1028 (%)	Unnamed Coal 1027 (%)	Unnamed Coal 1026 (%)	Unnamed Coal 1025 A (%)	Unnamed Coal 1025 B (%)
L. ovalis	19.0	10.0	14.5	25.0	29.5	11.0	3.5	10.5	33.0	4.5	6.5	3.5	6.0	18.5	33.5	5.5	5.0
L. striatus				X	0.5	0.5	X		1.0	X	0.5				X	X	0.5
L. vulgaris	3.0	0.5		0.5	0.5	X	X				X					X	
Renisporites confossus																X	
Latosporites minutus	1.5	2.0	1.0		3.5	0.5				1.0	1.0						
Punctatosporites minutus	1.5	4.0	11.5	6.5	6.0	5.0	0.5	7.5	10.0	7.5	3.0	1.5	5.0	4.5	2.5	4.0	5.5
Spinosporites exiguus	6.0	2.5	2.5	8.0	0.5			0.5	2.0		1.5		2.5				
Vestispora clara							X			0.5							
V. costata						X		X		X							
V. fenestrata						X	X		1.0	X				X	X		X
V. leavigata									X			X					
V. magna										X			X			X	
V. pseudo-reticulata	X		X	X									X		X	X	X
V. spp.			X		X												
Florinites mediapudens		2.5		0.5		5.0	1.0		10.0					0.5		1.5	3.0
F. millotti		X			X					X	X			0.5			
F. similis	6.5			1.5	2.0	1.5											
F. visendus		X		1.0	X									2.0	0.5		
F. volans	7.0	1.0	1.5	7.0	5.5					14.5	13.0	4.5	9.0	5.5	20.0		
F. spp.	12.0														17.5		
Wilsonites delicatus				0.5	0.5	1.0											
W. vesicatus	2.0	3.0		1.0	1.0		X				1.0	1.0	1.0		0.5	1.5	
Vesicaspora wilsonii																	0.5

*Formal systematics of taxa given in Appendix A.
†X = present but not in count.

TABLE 5. SPORE ANALYSIS OF THE GENTRY COAL BED (FORMERLY BATTERY ROCK COAL MEMBER) FROM ITS TYPE SECTION IN HARDIN COUNTY (MACERATION 587) AND OF THE "BATTERY ROCK" COAL (MACERATION 2461) FROM UNION COUNTY, KENTUCKY

Spore Taxa*	Macerations	
	587 (%)	2461 (%)
Deltoidospora sp.		0.5
Punctatisporites sp.		X†
Calamospora hartungiana		0.5
C. pedata		X
Granulatisporites granularis		0.1
G. microgranifer	7.5	
G. pallidus	X	
G. sp.		0.5
Cyclogranisporites minutus		X
Verrucosisporites sp.		0.5
Lophotriletes sp.	X	
Waltzispora prisca		0.5
Procoronaspora dumosa	X	
Apiculatasporites spinososaetosus		0.5
A. spinulistratus		X
A. variocorneus		X
A. sp.		X
Raistrickia abdita		X
R. prisca	X	
Microreticulatisporites harrisonii		0.5
Secarisporites remotus		X
Dictyotriletes sp.		0.5
Camptotriletes sp.		X
Ahrensisporites guerickei var. *ornatus*		X
Knoxisporites triradiatus		1.0
K. stephanephorus		X
Reticulatisporites splendens	X	
Savitrisporites nux		1.0
Grumosisporites varioreticulatus		X
Crassispora kosankei	X	
Densosporites annulatus		3.0
D. ruhus	X	
D. sinuosus	X	
D. sphaerotriangularis	X	
D. triangularis		2.0
D. spp.	0.5	1.5
Lycospora granulata	7.0	40.0
L. micropapillata		4.0
L. noctuina	2.5	
L. orbicula	7.5	
L. pellucida	70.0	27.0
L. pusilla		3.0
L. sp.		1.5
Cristatisporites indignabundus	X	3.0
Radiizonates striatus	0.5	
Schulzospora rara	4.0	1.5
Laevigatosporites desmoinesensis		2.0
L. ovalis		2.0
Florinites mediapudens		0.5
F. similis	0.5	
F. spp.		2.0

*Formal systematics of taxa given in Appendix A.
†X = present but not in count.

tains *Florinites mediapudens, Schulzospora,* and ~4% *Laevigatosporites.* Other differences in the spore assemblages can be seen in Table 5. The sample of the "Battery Rock coal" of western Kentucky is from an outcrop along the east bank of the Ohio River only about 2 mi (1.2 km) southeast of the type locality of the Gentry Coal.

Early reports indicated difficulties in correlating coals in the Caseyville Formation. Lesquereux (1857) and Glenn (1912) thought that the Bell coal bed of western Kentucky was equivalent to the Gentry Coal of Illinois, but Lee (1916) showed that the Gentry Coal occurs between conglomeratic sandstones (Pounds and Battery Rock Sandstones) in the Caseyville Formation and is considerably older than the Bell coal. Lee (1916, p. 16) noted the presence of a thin coal about 90 ft (27.4 m) below the top of the Pounds Sandstone and 175 ft (53.3 m) above the Gentry Coal on the Illinois side of the Ohio River and a coal at about the same position in Kentucky. This may be the coal that was mapped as the "Battery Rock coal" in Kentucky, because that coal was mapped as being about 120 ft (36.6 m) below the top of the Caseyville Formation. Kehn (1974) indicated that a lower coal occurs below the coal identified as the "Battery Rock coal," or about 250 ft (75.2 m) below the Bell coal. This lower coal might be equivalent to the Gentry Coal.

Phillips and Peppers (1984) compared the palynology of coals in northeastern Tennessee and the Illinois basin. Combined with additional data from eastern Kentucky coals, these comparisons helped verify correlations between the Illinois basin and the central part of the Appalachian coal region (Phillips et al., 1985). Correlations of the eastern Kentucky and Tennessee coals shown in Figure 3 and Plate 3 also correspond to correlations by Rice and Smith (1980). The names of the coals listed in Figure 3 and Plate 3 are vertically spaced according to what is thought to be their relative ages in the three states; therefore, there may appear to be gaps in the spore sequences. Most of the coal beds in eastern Kentucky and Tennessee discussed in this study (Table 4) are represented by only one sample; more detailed palynological studies are needed. Therefore, I did not attempt to extend the spore assemblage zonation proposed for the Illinois basin (Peppers, 1984) to the Appalachian coal region.

Rice (1978) and Rice et al. (1979b) correlated the Lower-Middle Pennsylvanian boundary with the middle of the Caseyville Formation largely on the basis of the last Illinois occurrence of *Schulzospora rara,* which earlier studies (Kosanke, 1950) had placed at the Gentry Coal. This correlation did not consider the fact that *S. rara* does not disappear at the Lower-Middle Pennsylvanian boundary but extends above the boundary in West Virginia. Peppers (1984) found that the range of *S. rara* extends to the Reynoldsburg Coal just above the top of the Caseyville Formation: thus, the correlation of Rice et al. (1979b) was correct after all. The continuous ranges of *Laevigatosporites* and *Florinites mediapudens* begin between the Gentry Coal and the "Battery Rock coal" of western Kentucky, and both species are already common in the Reynoldsburg Coal.[1] Thus, the Lower-Middle Pennsylvanian boundary in Illinois is between the Gentry and Reynoldsburg Coals and is correlated with the base of the Pounds Sandstone, as determined from comparison of ranges of *Schulzospora rara, Laevigatosporites, Florinites mediapudens,* and *Endosporites zonalis* in the Pennsylvanian stratotype area and the Illinois basin.

Western Interior coal province. The Lower-Middle Pennsylvanian Series boundary, as proposed by the USGS, is correlated with the base of the Dye Shale Member in the middle of the Bloyd Formation in Arkansas.

The position of the Lower-Middle Pennsylvanian boundary in the Western Interior Province is difficult to determine on the basis of palynology because of the scarcity of coal beds and the small number of palynological studies for that part of the Pennsylvanian. Although the boundary is supposed to correspond to the Morrowan-Atokan boundary (Bradley, 1956) in the Midcontinent, Arndt (1979, p. 73) noted that "recent studies in West Virginia and Kentucky have shown that the boundary between the Lower and Middle Pennsylvanian Series in the Appalachians is stratigraphically lower than the Morrowan-Atokan boundary." Henry and Gordon (1979) concluded that marine invertebrate faunas indicate that the upper part of the Morrowan is Middle Pennsylvanian, and Gordon (1984) stated that the USGS correlates the Lower-Middle Pennsylvanian boundary approximately with the base of the Dye Shale Member (Plate 1). The few pertinent palynological studies in the Western Interior coal province support this position.

The Bloyd Formation in the upper Morrowan of Arkansas is in floral zone 6 (*Neuropteris tennesseeana* and *Mariopteris pygmaea*) of Read and Mamay (1964). Wanless (1939, p. 36) stated that "David White, C. B. Read, and others have correlated the flora of the roof shale of the Battery Rock coal with that of the roof shales of the Sharon coal of Ohio, the Sewell coal of West Virginia, and the Lower Seaboard coal of Virginia, and with the shales in the Baldwin coal group, Morrowan formation of Arkansas."

Coal beds are not known in the Lower Pennsylvanian of Oklahoma. Wilson (1965) illustrated spores from Morrowan shale and sandstone in Ti Valley, but some of them had been redeposited from older rocks. Palynological correlation of the Morrowan with other regions is therefore difficult. Recycling of spores in Morrowan and Atokan rocks in the Ouachita Mountains was also discussed by Wilson (1976b).

Loboziak et al. (1984) reported on the palynology of six samples from the Hale and Bloyd Formations in northern Arkansas. Included was one coal sample from the Baldwin coal bed at the Evansville Mountain reference section of the Mor-

rowan (stop 9 of Sutherland, 1979). They recorded 17 taxa in the Woolsey Member, including *Punctatosporites minutus,* which has not been reported in coals of that age in the Illinois basin or the Appalachian coal region. The specimen illustrated as *P. minutus* by Loboziak et al. (1984) is difficult to interpret because it is not as elliptical as is typical for the species. They concluded that the Baldwin coal is early Westphalian A in age (Early Pennsylvanian). Dawson (1989) reported the presence of *Endosporites globiformis* in the Baldwin coal; however, Loboziak et al. (1984) and I did not observe that species, which first appears at the base of the Atokan and Westphalian B elsewhere.

The samples of the Baldwin coal that I studied (Table 2) have a low diversity of spores. Although maceration 2793 is from the same site sampled by Loboziak et al. (1984), our lists of palynomorphs are considerably different; in fact, *Savitrisporites nux* and *Laevigatosporites vulgaris* are the only species in common between their samples and mine. In maceration 2793, I observed abundant specimens of *Laevigatosporites desmoinesensis, Schulzospora rara,* and *Densosporites irregularis.* The other sample of the Baldwin coal that I studied (sampled in 1965, maceration 1443) contains rare specimens of *L. desmoinesensis,* and *D. irregularis* was not observed. Because the first appearance of a fossil is generally most reliable for making correlations, I consider that the spore assemblages in the Baldwin coal are only slightly younger than those in the Gentry Coal in Illinois. The Baldwin coal is thus in the upper part of the Lower Pennsylvanian, and the Lower-Middle Pennsylvanian boundary is at about the base of the Dye Shale as proposed by Gordon (1984).

The upper range of *Densosporites irregularis* is variable and extends into younger strata in the western coal regions than in the Appalachian coal region. In the Appalachian region, it disappears just after *Laevigatosporites* appears (Kosanke, 1982, 1988b; Eble and Gillespie, 1989). *Densosporites irregularis* extends from the Mississippian into some unnamed coals in the lower part of the Caseyville Formation in the southeastern part of the Illinois basin, but it disappears before *Laevigatosporites* appears. *Densosporites irregularis* extends above the first appearance of *Laevigatosporites desmoinesensis,* however, in northwestern Illinois (Peppers, unpublished data). Rare specimens of *Laevigatosporites desmoinesensis* were recorded by Ravn and Fitzgerald (1982) in the Westphalian A Wildcat Den Coal in eastern Iowa. *Densosporites irregularis* also occurs in the Wildcat Den Coal and ranges as high as the Blackoak Coal (Ravn, 1979). However, *D. irregularis* hasn't been reported as high as in the Pope Creek Coal in Illinois, which is approximately equivalent to the Blackoak Coal. Ravn divided the Blackoak Coal into 86 units for detailed palynological study. Because of this high level of detail, observations were made of very rare taxa, which otherwise probably would have been missed in one channel sample of coal. Another possibility is that the specimens were redeposited from older strata. The Wildcat Den Coal may represent a peat formed in a well-drained envi-

---

[1]As mentioned in the previous discussion, the Lower–Middle Pennsylvanian boundary in the Appalachian region is between the first appearance of *Laevigatosporites* below and *Florinites* above. *Endosporites zonalis* also begins its range just above the boundary in the Appalachian region and between the "Battery Rock coal" of western Kentucky and the Reynoldsburg Coal in the Illinois basin.

ronment along the edge of the Illinois basin, and the earlier appearance of *Laevigatosporites* in that coal may correspond more closely to its first appearance in western Europe (Clayton et al., 1977), where noncoal strata are also included in determining spore ranges.

Western Europe. In this study, the Lower-Middle Pennsylvanian boundary, as proposed by the USGS, is correlated with the middle of the Westphalian A and the base of the *Radiizonates aligerens* spore assemblage zone (Clayton et al., 1977) in western Europe.

In Great Britain, the Lower-Middle Pennsylvanian boundary is correlated with the middle of the *Radiizonates aligerens* assemblage VI, which is in the middle of the Lower Coal Measures (Smith and Butterworth, 1967). The base of assemblage VI is a little below the middle of the Westphalian A, but Clayton et al. (1977) placed the base of their *Radiizonates aligerens* (RA) zone at the middle of the Westphalian A. According to Owens et al. (1978), the base of the RA Zone of Clayton et al. corresponds to the base of the *Radiizonates aligerens* assemblage VI of Smith and Butterworth.

The base of the *R. aligerens* assemblage VI marks the beginning of the ranges of *R. aligerens, Endosporites zonalis, Densosporites sphaerotriangularis,* and *Florinites mediapudens,* as well as the continuous range of *Laevigatosporites* in Britain. Clayton et al. (1977) indicated, on the basis of rare occurrences at a few localities, that the ranges of the latter two taxa begin in the Namurian A but become noticeably more common just below the middle of the Westphalian A. In the Appalachian coal region, *Laevigatosporites* becomes more common just below the Lower-Middle Pennsylvanian boundary, and *F. mediapudens* becomes more common at the boundary. *Laevigatosporites* appears in one to several coals earlier than *Florinites mediapudens* in 14 of 17 of the individual coal fields in Great Britain where that part of the sequence occurs (Smith and Butterworth, 1967). In the other three coal fields, they appear at the same time. The appearance of *F. mediapudens* soon after that of *Laevigatosporites* also occurs in eastern Kentucky, Tennessee (Fig. 3), and the Illinois basin (Peppers, 1984).

The Lower-Middle Pennsylvanian boundary is correlated with the middle of the Modeste beds in the lower part of the Vicoigne Formation in the Nord-Pas-de-Calais coal basin of northern France studied by Loboziak (1971). Loboziak reached this conclusion because *Laevigatosporites* appears at the base of the formation and *Florinites mediapudens* begins its range slightly later. *Schulzospora* was not recorded in the western part of the basin, but Coquel (1976) reported that it extends almost to the top of the Vicoigne Formation (top of Westphalian A) in the eastern part of the basin.

Donets basin. I am correlating the Lower-Middle Pennsylvanian boundary (USGS) with the top of suite $C_2{}^2$ at Limestone H_1 in the Donets basin.

Owens et al. (1978) and Owens (1984) stated that *Laevigatosporites* increases to significant numbers at the base of the *Radiizonates aligerens* assemblage VI in Great Britain, which

is equivalent to the upper part of suite $C_2{}^2$. This level of increased occurrence of *Laevigatosporites* in coal and other rocks probably corresponds to its first appearance in the upper part of the Lower Pennsylvanian in the Appalachian coal region and the beginning of its continuous range in Great Britain. *Schulzospora rara* extends to the middle of suite $C_2{}^3$ and to the lower part of the Middle Pennsylvanian in the Appalachian coal region. *Florinites mediapudens* has not been reported in the Donets basin.

Westphalian A-B (Langsettian-Duckmantian) boundary

Western Europe. The Westphalian A-B boundary is well defined palynologically. According to Clayton et al. (1977), it is between the *Schulzospora rara* (VII) and *Dictyotriletes bireticulatus* (VIII) assemblages in Britain (Smith and Butterworth, 1967) and the *Radiizonates aligerens* (RA)[2] and overlying *Microreticulatisporites nobilis-Florinites junior* (NJ) spore assemblage zones in western Europe. The boundary is also at the base of the *Lonchopteris rugosa-Alethopteris urophylla* plant compression zone (Wagner, 1984) in Europe. *Lonchopteris* is common in Canada, but it has not been authenticated elsewhere in North America (Pfefferkorn and Gillespie, 1980). The Westphalian A-B boundary is at the base of the Middle Coal Measures in Great Britain (Smith and Butterworth, 1967), the Katherina Marine Band at the base of the Hendrick Formation in The Netherlands (Van Wijhe and Bless, 1974), and at the base of the Essener Schichten in the Ruhr basin (Grebe, 1972). It is also at the Quaregnon Marine Band in Belgium (Somers, 1971) and at the Poissonnière Marine Band at the base of the D'anzin Formation in Nord-Pas-de-Calais basin in France (Loboziak, 1971).

The index spores *Schulzospora rara, Radiizonates striatus, R. aligerens,* and *Lycospora noctuina* disappeared near the Westphalian A-B boundary (Plate 2). *Schulzospora rara* was not reported, however, by Van Wijhe and Bless (1974) or Van de Laar and Fermont (1989) in Westphalian A strata in The Netherlands. *Sinuspores sinuatus* disappeared, and *Laevigatosporites* and *Florinites* became more abundant soon after the end of the Westphalian A. *Endosporites globiformis* and *Microreticulatisporites nobilis* appeared at the Westphalian A-B boundary in Great Britain (Smith and Butterworth, 1967) and Belgium (Somers, 1971). In The Netherlands, *E. globiformis* appeared at the boundary, but *M. nobilis* first occurred in the lower part of Westphalian B (Van Wijhe and Bless, 1974; Van de Laar and Fermont, 1989). *Microreticulatisporites nobilis* also appeared at the boundary in northern France (Loboziak, 1971; Coquel, 1976). The range of *Vestispora pseudoreticulata* begins just above the Westphalian A-B boundary in Great Britain, The Netherlands, and Belgium. In northern France, its range begins in the Westphalian A. *Punctatosporites minutus* appears at the Westphalian A-B boundary in Great Britain and northern France (Loboziak, 1971), but it may

---

[2]According to Smith and Butterworth (1967), Grebe (1972), and Coquel (1976), however, *R. aligerens* doesn't quite reach the top of Westphalian A.

occur rarely in the upper Westphalian A in other parts of western Europe (Clayton et al., 1977).

Illinois basin. The boundary between the Westphalian A and B Stages is herein correlated with the top of the Reynoldsburg Coal, at the base of the Tradewater Formation in Illinois. The boundary is just below the No. 1b Bell coal near the base of the Tradewater Formation in western Kentucky and a little below the St. Meinrad Coal in the Mansfield Formation of Indiana. The Westphalian A-B boundary is about the same age as the Morrowan-Atokan boundary.

The Westphalian A-B boundary has been correlated, on the basis of palynological studies, with the top of the Caseyville Formation (Cross and Schemel, 1952; Bharadwaj, 1960; Hacquebard et al., 1960; Peppers and Popp, 1979; and Phillips and Peppers, 1984). Shaver and Smith (1974) and Shaver (1984) correlated the boundary, by use of fusulinids and ostracods, with about the position of the Tarter Coal Member at the middle of the Abbott Formation (now called lower half of the Tradewater Formation) and above the Morrowan-Atokan boundary. Peppers (1984) and Phillips et al. (1985) correlated the Westphalian A-B boundary with the top of the Reynoldsburg Coal in Illinois at the base of the Tradewater Formation.

The Westphalian A-B boundary is at the boundary between the *Schulzospora rara-Laevigatosporites desmoinesensis* and *Micro-reticulatisporites nobilis-Endosporites globiformis* spore assemblage zones in the Illinois basin (Peppers, 1984). Although the latter assemblage zone correlates with the *Microreticulatisporites nobilis-Florinites junior* spore assemblage zone in Europe, *F. junior* generally has not been reported in North America.

Schulzospora rara last appears in the Reynoldsburg Coal, in which it is rare, and at the top of the Westphalian A in Europe. *Radiizonates striatus* disappears just below the coal (Fig. 3); whereas, *Lycospora noctuina* and *Sinuspores sinuatus* disappear a little above the coal and at or a little above the Westphalian A-B boundary in Europe. In Illinois, *Endosporites globiformis* begins its range in an unnamed coal, which is of local extent (near Makanda in Jackson County) and slightly younger than the Reynoldsburg Coal. It also occurs in the overlying No. 1b Bell coal (western Kentucky) and St. Meinrad Coal (Indiana), which are correlative (Peppers and Popp, 1979), as well as at the base of the Westphalian B. No coal equivalent to the Reynoldsburg Coal has been identified in Kentucky or Indiana. *Microreticulatisporites nobilis* and *Vestispora pseudoreticulata* appear just above the Reynoldsburg Coal and the Westphalian A-B boundary in Europe, except in northern France (Loboziak, 1971; Coquel, 1976), where *V. pseudoreticulata* was reported in Westphalian A strata. *Lophotriletes gibbosus* first appears in the Reynoldsburg Coal, in the lower Westphalian B in northern France (Coquel, 1976), and in the upper Westphalian A in the Ruhr basin (Grebe, 1972). Rare specimens of *Punctatosporites minutus* appear for the first time in the lower part of Westphalian B in Illinois and at the middle Westphalian A in Europe. The range of *Vestispora magna* begins just above the Reynoldsburg Coal, and at the Westphalian A-B boundary in Great Britain and northern France (Coquel, 1976). *Punctatisporites minutus, Spackmanites habibii,* and *Renisporites confossus* appear just above the Reynoldsburg Coal, and *Laevigatosporites striatus* appears in the coal. These species have not been reported in Europe, but they may be used for making correlations between coal basins in the United States. Note the slight difference in the spelling of *Punctatosporites*, which has one long suture, and *Punctatisporites*, which has a trilete suture.

The name Bell Coal was used for the first time in Illinois by Peppers and Popp (1979), but the name was not formally introduced for usage in Illinois. The name Tunnel Hill Coal Bed was introduced for a coal that was mapped recently in Johnson County, Illinois, and was correlated palynologically with the Bell Coal (Trask and Jacobson, 1990). The USGS uses the name No. 1b (Bell) coal bed (Kehn, 1974), and the Kentucky Geological Survey uses the name No. 1b Bell coal (Williams et al., 1982). Owen (1855) introduced the name as "Place of Bell Coal" for the coal at Bell's Mine. Kosanke et al. (1960) correlated the Bell coal with the Willis Coal Member in Illinois, but palynological studies demonstrate that the Bell coal is much older than the Willis and slightly younger than the Reynoldsburg Coal (Peppers, 1977). Mapping indicates that the Tunnel Hill Coal occurs 10 to 20 m (32.8 to 65.6 ft) above the Reynoldsburg Coal (Trask and Jacobson, 1990). The Bell coal is equivalent to the Hawesville coal in the northern part of western Kentucky (unpublished data of Peppers) and the St. Meinrad Coal in Indiana. Table 6 compares the palynological assemblages in the No. 1b Bell coal near Bell Mines Church, Union County, Kentucky, with the Tunnel Hill Coal in Johnson County, Illinois.

Western Interior coal province. The Westphalian A-B boundary correlates with the base of the Kilbourn Formation in Iowa, the top of the McLouth Formation in northwestern Missouri, the base of the Trace Creek Shale Member in Arkansas, and the base of the Atoka Formation in Oklahoma. Thus, it is equivalent to the Morrowan-Atokan boundary, which is discussed later.

Manger and Sutherland (1984) correlated the Westphalian A-B boundary with the base of the Trace Creek Shale in Arkansas with the understanding that the Trace Creek is considered the lower part of the Atoka Formation (as recommended by Sutherland and Grayson, 1978), rather than the upper part of the Bloyd Formation, as has been done in the past. An Atokan age for the Trace Creek Shale was indicated by conodont evidence (Grayson and Sutherland, 1977). Zimbrick (1978) concluded that the shale is a lateral facies of the Atoka Formation in Oklahoma. Phillips and Peppers (1984) and Phillips et al. (1985) correlated the Westphalian A-B boundary with the middle of the Morrowan and the upper part of the Morrowan, respectively. The boundary was considered to be equivalent to the middle of the Atokan by Harland et al. (1990). According to Ross and Ross (1985, 1988), who correlated transgressive-regressive sequences on the basis of foraminiferal zonation, the Westphalian A-B boundary is equivalent to the early Desmoinesian in the Midcontinent.

TABLE 6. SPORE ANALYSIS OF THE BELL COAL BED NEAR BELL MINES, UNION COUNTY, KENTUCKY (MACERATION 2255), AND THE TUNNEL HILL COAL BED IN JOHNSON COUNTY (MACERATION 2611) AND POPE COUNTY (MACERATION 2978), ILLINOIS

Spore taxa*	Macerations			Spore taxa*	Macerations		
	2255 (%)	2611 (%)	2978 (%)		2255 (%)	2611 (%)	2978 (%)
Deltoidospora levis			X†	*Knoxisporites stephanephorus*			X
D. priddyi	X			*K. triradiatus*		X	
D. sphaerotriangula		X		*K.* sp.			X
				Reticulatisporites muricatus	X		
Punctatisporites glaber			0.5	*R. polygonalis*			0.5
P. irrasus	X	X		*R.* sp.	X		X
P. minutus	X		X	*Savitrisporites nux*	X	X	
				Grumosisporites varioreticulatus	0.5		X
Calamospora breviradiata		0.5					
C. hartungiana	0.5			*Crassispora kosankei*	1.0		2.5
C. liquida		X		*Granasporites medius*	X		X
C. mutabilis		X					
C. straminea				*Densosporites annulatus*	X	X	0.5
C. spp.	X		0.5	*D. glandulosus*	X		
				D. sphaerotriangularis	0.5	X	3.0
Granulatisporites adnatoides		X	0.5	*D. triangularis*			X
G. granularis	X	X					
G. granulatus	X			*Lycospora granulata*	43.5	53.0	22.5
G. minutus		1.0		*L. micropapillata*	2.5	X	X
G. pallidus	X			*L. noctuina*			1.5
G. piroformis			X	*L. pellucida*	39.0	31.0	49.5
G. verrucosus	0.5			*L. pusilla*	4.5	10.0	10.0
G. sp.		X		*L. rotunda*	X	X	
				L. subjuga	X		
Cyclogranisporites cf. *aureus*	X						
C. minutus	X		1.0	*Cristatisporites indignabundus*	X		X
C. sp.			0.5	*Cirratriradites saturnii*	X		
Sinuspores sinuatus	X	X	X	*Endosporites globiformis*	4.5	1.5	X
Converrucosisporites sp.		X		*Alatisporites* sp.			X
				Hymenospora multirugosa		X	X
Verrucosisporites microtuberosus	X		X				
Lophotriletes gibbosus	X	X		*Laevigatosporites desmoinesensis*	X		3.5
L. microsaetosus		X		*L. medius*	X		
Anapiculatisporites minor	X			*L. ovalis*	2.0	2.0	
				L. striatus	X		
Pustulatisporites crenatus	X		X				
Apiculatasporites spinososaetosus		X		*Renisporites confossus*	X	X	X
A. variocorneus	X		X	*Dictyomonolites swadei*		X	
A. sp.			1.5	*Vestispora costata*	X	X	
Planisporites granifer	X		0.5	*V. irregularis*			X
				V. pseudoreticulata		X	
Raistrickia breveminens	X	X					
R. crocea	X	X		*Florinites mediapudens*	0.5	X	
				F. similis		1.0	
Spackmanites habibii	X			*F. volans*			1.5
Convolutispora florida		X	X	*Wilsonites circularis*	X	X	X
C. sp.	X		X	*W. delicatus*		X	
Microreticulatisporites concavus		X		*W. vesicatus*	0.5		
M. harrisonii		X	X	*Quasillinites diversiformis*		X	X
Dictyotriletes bireticulatus	X	X	X	*Trihyphaecites triangulatus*		X	
Ahrensisporites guerickei var. *ornatus*	X	X	X				
Triquitrites bransonii		X	X				
Reinschospora triangularis	X						

*Formal systematics of taxa given in Appendix A.
†X = present but not in count.

As is discussed in the section on the Lower-Middle Pennsylvanian boundary, Loboziak et al. (1984) studied the palynology of several samples from the Hale and Bloyd Formations in Arkansas and correlated the Westphalian A-B boundary with the upper part of the Trace Creek, which they considered late Morrowan in age. They concluded that the bottom of the member contains a middle Westphalian A spore assemblage, but their evidence is weak because of the small number of taxa found. *Schulzospora* occurs in the Woolsey Member but not at the top of the Dye Shale or in the overlying Trace Creek Shale. The top of the Trace Creek, which they considered the base of the Westphalian B, contains *Alatisporites pustulatus, Endosporites globiformis,* and possibly *Florinites junior,* all of which appear in the early part of Westphalian B in most regions. The Westphalian A-B boundary could just as well be near the base of the shale rather than at the top, because no samples were studied between the base of the Trace Creek Shale, which is Westphalian A in age, and the top of the shale. I am therefore correlating the Westphalian A-B boundary with the base of the Trace Creek in order to conform more closely with evidence from ammonoid studies (Saunders et al., 1977).

Ravn (1986) correlated the Westphalian A-B boundary with a hiatus between the top of the Caseyville Formation and the base of the Kilbourn Formation in Iowa, where *Schulzospora rara* disappears; I agree with this correlation. *Sinuspores sinuatus* extends into the lower part of the Kilbourn Formation in Iowa (Ravn, 1986), and to the top of the Westphalian A or lower part of the Westphalian B in western Europe. *Vestispora pseudoreticulata* and *Endosporites globiformis* first appear in a shale at the base of the Kilbourn Formation and at the beginning of Westphalian B in Europe, except in northern France, where *V. pseudoreticulata* appears in Westphalian A. Ravn's first record of *Microreticulatisporites nobilis* is in the Blackoak Coal of late Westphalian C age (late Westphalian B according to Ravn, 1986), and it also occurs in the upper part of the Riverton Formation in northwestern Missouri (Peppers et al., 1993). In Illinois and western Europe, it appeared considerably earlier, at the beginning of Westphalian B (Plate 2). *Punctatosporites minutus* appears just above the Westphalian A-B and Morrowan-Atokan boundaries in the Midcontinent and Illinois basin.

Appalachian coal region. The Westphalian A-B boundary in eastern Kentucky is correlated with the top of the Gray Hawk coal in the lower part of the Breathitt Formation west of the Pine Mountain fault (Plate 1) and with the top of the Splitseam coal east of the escarpment. The boundary is at the top of the Lower War Eagle coal in the lower part of the Kanawha Formation in West Virginia, at the top of the Hooper coal in Tennessee, and at the base of the Quakertown No. 2 coal in Ohio.

The boundary between the Westphalian A and B has been correlated with the top of the New River Formation (Cross and Schemel, 1952; Bharadwaj, 1960, Phillips, 1981; and Phillips and Peppers, 1984). Using a study of plant megafossils from the proposed stratotype of the Pennsylvanian System, Gillespie and Pfefferkorn (1979) correlated the A-B boundary with just above the Sewell coal bed in the New River Formation.

As mentioned previously, the last occurrence of *Schulzospora rara* (an indicator of the Westphalian A-B boundary) in West Virginia is in the Lower War Eagle coal (Table 3). Specimens comparable to *Endosporites globiformis,* which first appears at the Westphalian A-B boundary in Europe, were reported by Kosanke (1984) from the type outcrop of the Cedar Grove coal bed of West Virginia. He did not observe it in an older coal assigned with uncertainty to the Gilbert coal, but Kosanke (1988a) reported it along with *Schulzospora rara* from another sample he referred to as the Gilbert(?) coal. Another possible overlap of the stratigraphic ranges of *Schulzospora* and *Endosporites globiformis* (listed as *E. ornatus*) is in the Quakertown No. 2 coal in Ohio, which I correlate with the top of the Westphalian A (Kosanke, 1984). *Endosporites ornatus* was not recorded in the coal seam, but it was found in the seat rock and roof rock. *Punctatosporites minutus* also appears in the Gilbert(?) coal and at the Westphalian A-B boundary in most of Europe. The first appearance of *Vestispora* in West Virginia is represented by *Vestispora* cf. *V. costata,* also in the Gilbert(?) coal. *Vestispora costata* first appeared during late Westphalian B time in Europe, but *V. tortuosa,* which Ravn (1986) suggested is probably synonymous with *V. costata,* first occurred in late Westphalian A time in Europe. Because the identity of the Gilbert(?) coal is in doubt, its spore data are not included in the range charts.

I observed *E. globiformis* in the Mason coal of eastern Kentucky (Fig. 3), and it is common in the overlying Hance coal bed. *Vestispora pseudoreticulata* and *Laevigatosporites striatus* appear in the River Gem coal bed in Kentucky (Fig. 3), which is approximately correlative with the Hance coal. *Vestispora pseudoreticulata* appears just above the Westphalian A-B boundary in Europe.

In northeastern Tennessee, *Schulzospora rara* extends as high as the Morgan Springs coal bed below the Rockcastle Conglomerate, and *Endosporites globiformis* first occurs in the Poplar Creek coal bed (Phillips and Peppers, 1984). Neither species was found in the intervening Hooper coal bed (Table 4). *Punctatosporites minutus* appears later in Tennessee (Jellico coal) than in West Virginia. *Microreticulatisporites nobilis, Vestispora pseudoreticulata,* and *Laevigatosporites striatus* first appear in the Coal Creek coal bed in eastern Tennessee (Fig. 3). *Microreticulatisporites nobilis* and *V. pseudoreticulata* appear near the Westphalian A-B boundary in Europe. *Laevigatosporites striatus* was reported from the Westphalian B in northern England by Butterworth et al. (1988) and in the Stephanian in France by Alpern (1959). The species has not been extensively reported in Europe. *Punctatisporites minutus* first appears in the Blue Gem coal in Tennessee and in the Bell Coal in Illinois, both in the lower Westphalian B (Fig. 3). *Lycospora noctuina* disappeared at the Westphalian A-B boundary in Europe and in the lower part of the Breathitt and Kanawha Formations of the Appalachian coal regions. *Vestispora magna* appeared a little later in the Appalachian region and just above the base of the Westphalian B in Germany.

Donets basin. The Westphalian A-B boundary is correlated with Limestone H_4 in the middle of the C_2^3 suite (upper Bashkirian) in the Donets basin.

Teteryuk (1976) correlated the Westphalian A-B boundary with Limestone H_5 and the base of his *Endosporites globiformis-Bellispores bellus* spore assemblage zone in the Donets basin. He showed that *Schulzospora* disappeared and *Vestispora* and *Endosporites globiformis* appeared in coal near that limestone. Owens et al. (1978) indicated that *Punctatosporites minutus* also appeared at the beginning of the Westphalian B. Teteryuk later (1982) stated that the spore assemblages beginning at Limestone H_4 compare closely with those at the base of NJ spore assemblage zone because of the abundance of *Laevigatosporites* and *Florinites* and the appearance of *M. nobilis, Vestispora costata,* and *Florinites junior*. The base of the *Microreticulatisporites nobilis-Florinites junior* (NJ) spore assemblage zone (the base of the Westphalian B in western Europe) was also correlated with Limestone H_5 by Coquel et al. (1984), including Teteryuk. Because the 1984 paper was prepared for the Ninth International Carboniferous Congress (held in 1979) and was not published until almost 5 years later, the 1982 paper is probably a later interpretation. Fissunenko and Laveine (1984) correlated the Westphalian A-B boundary with the slightly younger Limestone I_2 in the C_2^4 suite on the basis of fossil plant compressions.

Morrowan-Atokan Stage boundary (Lower-Middle Pennsylvanian Series boundary of the Midcontinent)

Western Interior coal province. The term "Morrowan Series" was first used by Moore (1932) for the Morrow Formation, and Moore (1944) included it in the Midcontinent time rock divisions of the Pennsylvanian System (Morrowan, Lampasas, Desmoinesian, Missourian, and Virgilian). The Carboniferous Subcommittee of the American Association of Petroleum Geologists (Cheney et al., 1945) designated the reference section for the Morrowan. They correlated it with the New River Formation and the middle of the Pottsville Series in the Appalachian region. The Morrowan Series in Oklahoma was defined by Moore (1947) to include strata between the underlying Mississippian and the overlying Atoka Formation. Henbest (1962a, 1962b) formally defined the Morrowan Series by designating type sections in Washington County (Arkansas) for the formations within the series. The Morrowan is currently considered a stage in the Lower Pennsylvanian Series. I am correlating the Morrowan-Atokan boundary with the Westphalian A-B boundary; the palynological criteria for identifying the boundaries were discussed in the section on the Westphalian A-B boundary.

The stratigraphic sequence at Atoka, Oklahoma, named the "Atoka shale" by Taff and Adams (1900), is essentially unfossiliferous. The equivalent sequence in Texas was later described as the Lampasas Series (Cheney, 1940; Moore, 1944), in which the family Fusulinidae first appears. Thompson (1942) named the pre-Desmoinesian section in New Mexico the Derryan, but the

Morrowan was undifferentiated. The Atoka Formation was raised to series status by Spivey and Roberts (1946), and included strata between the top of the Morrowan Series and the base of the Hartshorne Formation in the overlying Desmoinesian Series. A disconformity at the base of the Atoka at its type area represents an erosional break (Sutherland and Grayson, 1978).

The problem of defining the Morrowan-Atokan boundary was discussed by Shaver and Smith (1974), Lane and Straka (1974), Dunn (1976), Thompson (1979), Shaver (1984), Grayson (1984), Lane and West (1984), Sutherland and Manger (1983, 1984), Manger and Sutherland (1984), Ravn (1986), Sutherland and Grayson (1992), and others. Most stratigraphers working in the Midcontinent use the terms Morrowan-Atokan and Lower-Middle Pennsylvanian boundaries interchangeably.

Sutherland and Manger (1983) summarized the difficulty of defining the Morrowan-Atokan boundary by stating that most groups of fossils occur sporadically and that a distinct biostratigraphic event did not occur at the boundary. Generally, the Atokan Stage has been considered the zone of *Profusulinella* and *Fusulinella* (Douglass and Nestell, 1984), but there is some question concerning the actual biostratigraphic range of *Profusulinella* in the Atoka Formation. Sutherland and Manager (1983) proposed that the base of the Atokan should be placed approximately at the first appearance of *Pseudostaffella* and *Eoschubertella*.

The Trace Creek Shale was transferred from the Morrowan Bloyd Formation to the base of the Atoka Formation (Sutherland and Grayson, 1978; Manger and Sutherland, 1984) because the Trace Creek is overlain by a disconformity and has been traced from Arkansas to Oklahoma, where it is the basal part of the Atoka Formation. This change makes the Morrowan-Atokan boundary in Oklahoma older than previously recognized.

Because strata originally assigned to the lower part of the Desmoinesian in Iowa correlate with the Atokan, Ravn et al. (1984) included the Atokan Series in the lower part of the Des Moines Supergroup.

In the Forest City Basin in northwestern Missouri and eastern Kansas, the Morrowan-Atokan boundary is at the top of the McLouth Sandstone Formation (Peppers et al., 1993). *Schulzospora rara* and *Densosporites irregularis* extend to the top of the formation, and *Endosporites globiformis* first appears just above the top of the formation. *Pilosisporites aculeolatus* appears a little later in an unnamed formation above the McLouth Formation. The beginning of the ranges of *Florinites mediapudens* and *Laevigatosporites desmoinesensis* occur in the lower part of the McLouth.

Illinois basin. Current palynological information indicates that the Morrowan-Atokan boundary is correlated with the Westphalian A-B boundary at the top of the Reynoldsburg Coal in Illinois, just below the No. 1 Bell coal in western Kentucky, and just below the St. Meinrad Coal in Indiana.

Previously, the Morrowan Stage was alternatively considered equivalent to only the lower half of the Caseyville Formation of Illinois (Wanless, 1956; Kosanke et al., 1960), most of the Caseyville Formation (Moore 1944), or the entire Caseyville (Willman et al., 1967; McKee and Crosby, 1975; Hopkins and Simon, 1975; Atherton and Palmer, 1979; Peppers and Popp, 1979; Phillips, 1981; Manger and Sutherland, 1984; Ravn, 1986). The Morrowan has also been correlated with the Caseyville Formation and the lower part of the Tradewater Formation of western Kentucky and Illinois (Rice, 1978; Rice et al., 1979b). Shaver and Smith (1974), Shaver (1984), and Shaver et al. (1985) placed the boundary considerably higher, at the base of the Brazil Formation, largely as a result of their ostracod studies and their interpretation of fusulinid studies. Shaver et al. (1985) extended this position to other areas in the Midwestern basins and arches region correlation chart for COSUNA. Phillips and Peppers (1984) followed the interpretation of Shaver and Smith (1974). Later, on the basis of new information from palynological studies (Jacobson et al., 1983; Phillips et al., 1985), they correlated the boundary with the lower part of the Abbott Formation (now the lower part of the Tradewater Formation) in Illinois.

Read and Mamay (1964) included the Atokan in their plant compression zones 7 (*Megalopteris*) and 8 (*Neuropteris tenuifolia*). Strata associated with the Tarter Coal of Illinois, however, were placed in floral zone 7 and described as early Atokan in age. Palynological evidence indicates that the Tarter Coal is late Atokan in age (Plate 1). Read and Mamay considered a flora from strata associated with the Eagle coal in the Kanawha Formation as characteristic of zone 8, which includes the major and upper part of the Atokan Stage. This would make the Eagle coal of West Virginia younger than the Tarter Coal of Illinois, but other evidence does not support this relation.

A recently discovered thin limestone in southern Illinois has yielded marine fauna, including the conodont *Idiognathoides ouachitensis*, that indicates a late Morrowan to early Atokan age (Jacobson et al., 1983). Although the conodont assemblage zone *Neognathodus-Idiognathoides ouachitensis* is late Morrowan in age (Grayson, 1984), *I. ouachitensis* extends well up into the Atokan in Oklahoma (Groves and Grayson, 1984; Grubbs, 1984) and Texas (Manger and Sutherland, 1984). A coal about 10 ft (3.1 m) below the limestone in section 19, T11S, R4E, Johnson County, Illinois (Table 7), had been identified as the Bell Coal and early Atokan by Jacobson et al. (1983) on the basis of palynological interpretation (Peppers, unpublished data, 1993). Palynological study of additional coal samples from the same area indicates that the coal is a little younger than the Bell Coal (Table 6), but older than Lee's (1916) Smith coal in western Kentucky. The Johnson County coal is biostratigraphically older than the first appearance of *Laevigatosporites globosus*, which is a little below the Smith coal. *Triquitrites sculptilis*, which appears earlier than *L. globosus* and above the Bell Coal, was also not observed. The coal in question therefore is only a little younger than the Bell Coal.

Palynological data used for identifying the Morrowan-Atokan boundary are the same as those for the Westphalian A-B boundary, which were given in the discussion of that boundary.

Appalachian coal region. The Morrowan-Atokan boundary is tentatively correlated with the top of the Gray Hawk coal bed in eastern Kentucky, at the top of the Hooper coal bed in Tennessee, at the base of the Quakertown No. 2 coal bed in Ohio, and at the top of the Lower War Eagle coal bed in West Virginia.

As stated previously, according to the recommendations of the USGS Geological Names Committee (Bradley, 1956), the Lower Pennsylvanian Series in the Appalachians corresponds approximately to the "Morrow" of the Midcontinent region. The Morrowan-Atokan boundary in the Midcontinent is considerably younger, however, than the Lower-Middle Pennsylvanian boundary of the Appalachian coal region (Gordon, 1984).

The Morrowan-Atokan boundary has been correlated with the base of the Corbin Conglomerate just above the top of the Lee Formation in eastern Kentucky (Moore, 1944), and at the top of the Lee and New River Formations (McKee and Crosby, 1975). A late Morrowan age was assigned to the Kendrick Shale Member by Furnish and Knapp (1966) using ammonoids and by Strimple and Knapp (1966) using crinoids. The Kendrick Shale is about middle Atokan in age, according to palynological evidence. Rice (1978) drew the top of the Morrowan in Kentucky between the Kendrick Shale (just below the Whitesburg coal bed) and Magoffin Members (a little above the Hamlin coal zone), as suggested by the paleontological studies. Henry and Gordon (1979) described a fauna from the Eagle Limestone of West Virginia, and they dated it as probably late Morrowan. The Eagle Limestone and coal are a little above the Lower War Eagle coal, which is correlated in this report with the base of the Atokan.

The spore taxa used to define the Morrowan-Atokan and Westphalian A-B boundaries in the Appalachian region are the same and were mentioned under the discussion of the latter boundary.

Western Europe. Palynological studies support correlation of the Morrowan-Atokan boundary with the Westphalian A-B boundary in Europe.

Gray (1979), Arkle et al. (1979), Einor et al. (1979), and Klein (1990) correlated the Morrowan-Atokan boundary with strata as low as the base of the Westphalian A, and stratigraphic distribution of bryozoans and foraminifers led Ross (1984) and Ross and Ross (1985, 1988) to support that position. Ammonoid zonation led Ramsbottom and Saunders (1984) to correlate the Morrowan-Atokan boundary with the Westphalian B-C boundary. Shaver and Smith (1974) used ostracod studies to correlate the Morrowan-Atokan boundary with the latest Westphalian B. Because the Trace Creek Shale in Arkansas has been transferred (Sutherland and Grayson, 1978) from the Morrowan to the Atokan, however, the Morrowan-Atokan boundary would be closer to the Westphalian A-B boundary. Shaver (1984) correlated the boundary with the youngest Westphalian A. Sohn and Jones (1984), who also used ostracod ranges, and Durden (1984a), who studied insect faunas, were essentially in agreement by considering the Morrowan-Atokan and Westphalian

TABLE 7. SPORE ANALYSIS OF THE UNNAMED COAL (MACERATION 2824) UNDERLYING THE LIMESTONE CONTAINING LATE MORROWAN OR EARLY ATOKAN CONODONTS IN JOHNSON COUNTY, ILLINOIS

Spore taxa*	Maceration 2824 (%)	Spore taxa*	Maceration 2824 (%)
Deltoidospora priddyi	X†	*Grumosisporites varioreticulatus*	0.5
Punctatisporites cf. *edgarensis*	X	*Crassispora kosankei*	X
P. flavus	X	*Granasporites medius*	1.0
P. obesus	X	*Densosporites annulatus*	X
P. minutus	2.5		
		Lycospora granulata	13.5
Calamospora breviradiata	0.5	*L. micropapillata*	11.5
C. hartungiana	1.5	*L. orbicula*	5.5
C. straminea	X	*L. pellucida*	20.5
		L. pusilla	8.0
Granulatisporites granulatus	0.5	*L. rotunda*	3.0
G. minutus	X		
G. pallidus	1.0	*Cristatisporites indignabundus*	X
G. piroformis	1.5	*Endosporites globiformis*	8.0
		Laevigatosporites desmoinesensis	8.5
Cyclogranisporites aureus	2.0	*L. medius*	X
C. leopoldi	0.5		
C. minutus	4.5	*Renisporites confossus*	X
Verrucosisporites microtuberosus	X	*Punctatosporites minutus*	0.5
		Vestispora cf. *fenestrata*	X
Lophotriletes gibbosus	X	*V. laevigata*	X
L. mosaicus	X	*V. pseudoreticulata*	X
L. pseudaculeatus	X		
L. rarispinosus	X	*Florinites mediapudens*	1.0
Apiculatasporites spinososaetosus	X	*F. millotti*	0.5
Pilosisporites williamsii	X	*F. volans*	X
Raistrickia crocea	X		
Spackmanites habibii	2.0	*Wilsonites delicatus*	X
Convolutispora cf. *florida*	X	*W. vesicatus*	0.5
C. sp.	0.5	*Potonieisporites elegans*	X
		Quasillinites diversiformis	X
Microreticulatisporites harrisonii	X	*Trihyphaecites triangulatus*	X
Ahrenisporites guerickei	X		
Triquitrites bransonii	0.5		
Knoxisporites triradiatus	X		
Savitrisporites nux	X		

*Formal systematics of taxa given in Appendix A.
†X = present but not in count.

A-B boundaries as equivalent. Moore (1944) and Manger and Sutherland (1984) also correlated the two boundaries with each other. Phillips et al. (1985) concluded, largely on the basis of studies that considered the Trace Creek Shale as Morrowan, that the Morrowan-Atokan boundary is equivalent to about middle Westphalian B.

Donets basin. I propose that the Morrowan-Atokan boundary (Westphalian A-B boundary) correlates with Limestone H₄ near the middle of the C₂³ suite (upper Bashkirian) in the Donets basin.

Bashkirian-Moscovian boundary

Donets basin. The type area of the Bashkirian Stage is in Bashkiria in the southern Urals, and the type area of the Moscovian Stage is in the Moscow basin on the Russian plat-

form. In the Donets basin, which contains the most complete section of the Carboniferous in the former Soviet Union, the Bashkirian-Moscovian boundary is at the base of Limestone K₃ in suite C₂⁵. Teteryuk (1982) stated that the upper Bashkirian and lower Moscovian palynological assemblages are characterized by an abundance of *Lycospora*, *Laevigatosporites*, and *Florinites*. The upper Bashkirian includes the appearance of *Vestispora costata*, *Microreticulatisporites nobilis*, and *Florinites junior*, as well as the disappearance of *Grumosisporites varioreticulatus* and *Savitrisporites nux*. The lower Moscovian marks the first appearance of *Torispora securis*, *Triquitrites sculptilis*, *Murospora kosankei*, and *Vestispora fenestrata*.

Western Europe. The boundary between the Bashkirian and Moscovian Stages is correlated in this study with the late Westphalian B of western Europe.

The position of the Bashkirian-Moscovian boundary in the

Donets basin, with respect to European chronostratigraphy, has been the subject of controversy. Soviet stratigraphers generally correlated the Bashkirian-Moscovian boundary in the Moscow and Donets basins with the Westphalian B-C boundary (Stepanov et al., 1962; Teteryuk, 1976, 1982; Bouroz et al., 1978; Einor et al., 1979; Solovieva et al., 1984). On the basis of ostracod and fusulinid studies, Shaver and Smith (1974) correlated the boundary with the upper part of the Westphalian A. In their geologic time scale, Harland et al. (1990) made the same correlation.

Owens et al. (1978) compared spore assemblage zones in Dinantian to Westphalian deposits of western Europe and equivalent strata in the Donets basin. They concluded that broad correlations between the two regions could be made, but definitive correlations of zones were not possible at that time. Using spores, Teteryuk (1976) correlated the Bashkirian-Moscovian boundary in the Donets basin with the earliest Westphalian C, and he lowered it to the Westphalian B-C boundary in 1982. Wagner and Higgins (1979) thought it unlikely that the Bashkirian-Moscovian and Westphalian B-C boundary would coincide because there is no floral break at the Westphalian B-C boundary (Aegir/Mansfield Marine Band). Van Ginkel (1965), Wagner and Higgins (1979), Wagner and Bowman (1983), and others proposed, on the basis of paleobotanical and paleontological evidence from Spain, that the Bashkirian-Moscovian boundary lies within the Westphalian A. They also recognized that a higher position, in the Westphalian C or at the Westphalian B-C boundary, is indicated by the data from the Donbass basin. The absence of *Lyginopteris* from coal balls above the level of the k8 coal in the lower Moscovian of the Donets basin led Phillips (1981) to place the Bashkirian-Moscovian boundary in the upper Westphalian B, because that pteridosperm is well represented in Europe from the Visean into the Westphalian B. Phillips and Peppers (1984) and Phillips et al. (1985) also considered the Bashkirian-Moscovian boundary as late Westphalian B in age.

Owens et al. (1978) showed that *Punctatosporites minutus* begins its range in the upper Bashkirian in the Donets basin. In western Europe, it first appears in upper Westphalian A to lower Westphalian B strata. *Torispora* first appears in the lower Moscovian in the Donets basin at Limestone L_1 or L_3 (Inosova et al., 1975; Teteryuk, 1976, 1982). Owens et al. (1978) pointed out that the first appearance of *Torispora* in early Westphalian C time is a major palynological event that can be used for making broad correlations. *Vestispora laevigata* appears in the lower part of the Moscovian in the Donets basin (Teteryuk, 1976) and in the Westphalian C in Great Britain (Smith and Butterworth, 1967), but Coquel (1976) reported it from as low as the upper Westphalian A in northern France. Because this species can be confused with other species of *Vestispora* that are overmacerated or poorly preserved, this earlier reported occurrence may not be reliable.

Fissunenko and Laveine (1984) compared stratigraphic ranges of plant compressions and spores and found that the

Westphalian B-C boundary as established in France can be correlated with the $C_2{}^5$ Suite at Limestone K_7, which would place the Bashkirian-Moscovian boundary below the Westphalian B-C boundary. It could also be correlated at Limestone K_3, which would place the Bashkirian-Moscovian boundary at the Westphalian B-C boundary. The faunal correlation of the Bashkirian-Moscovian boundary in the Donbass sequence is well established, but further investigations of faunas and floras are needed before more precise correlations can be made in the Donets basin.

Illinois basin. In Illinois the Bashkirian-Moscovian boundary is correlated with a position a little below the middle of the lower half of the Tradewater Formation, between the Tunnel Hill Coal Bed and the Manley Coal Member. In Indiana the boundary is correlated with the base of the Pinnick Coal Member, and in western Kentucky it is correlated between the Bell and Smith coals.

Shaver and Smith (1974), and Shaver (1984), on the basis of fusulinid and ostracod studies, and Shaver et al. (1985, 1986) correlated the Bashkirian-Moscovian boundary with the top of the Mansfield Formation in Indiana and the middle of the Abbott Formation (now lower half of Tradewater Formation) in Illinois. Teteryuk (1982) used palynology to correlate the Bashkirian-Moscovian boundary with the boundary between the Abbott and Spoon Formations (now the middle of the Tradewater Formation) in Illinois. Phillips and Peppers (1984) indicated, by use of the coal-ball and palynological studies, that the Bashkirian-Moscovian boundary correlates with beds at about the middle of the lower half of the Tradewater Formation.

Triquitrites pulvinatus, T. tribullatus, T. sculptilis, Vestispora fenestrata, and *V. laevigata* first appear just above the base of the Moscovian (Teteryuk, 1976) in the Donets basin. *Triquitrites sculptilis* appears a little below the middle, *V. laevigata* appears at about the middle, and *T. pulvinatus* appears a little above the middle of the lower half of the Tradewater Formation in Illinois (Peppers, 1984). According to Inosova et al. (1975, 1976), *Torispora* and *Foveolatisporites fenestrata* (= *Vestispora fenestrata*) have ranges that begin in the lower one-third of the Moscovian. These species occur for the first time in the Illinois basin at about the Smith coal of Lee (1916) in western Kentucky, or a little above the middle of the lower half of the Tradewater Formation (Plate 2).

The Bashkirian-Moscovian boundary is just below the Pinnick Coal, which overlies the "Hindostan Whetstone Beds" in southern Indiana. The boundary is intermediate in age between the St. Meinrad Coal and Blue Creek Coal. The Pinnick Coal marks the appearance of *Laevigatosporites globosus* (Table 1).

Western Interior coal province. The Bashkirian-Moscovian boundary is correlated with the lower part of the Kilbourn Formation in Iowa and a little below the middle of an unnamed formation between the McLouth Formation and Warner sandstone in northwestern Missouri.

In Iowa, *Triquitrites sculptilis, Vestispora fenestrata,* and *V. laevigata* first appear in unnamed coals in the Kilbourn Formation (Ravn, 1986) and in early Moscovian time. *Punctatosporites minutus* also begins its range in the Kilbourn Formation and during late Bashkirian time. Because coals in the Kilbourn occur throughout the formation, are of limited lateral extent, and were not differentiated by Ravn (1986), a more precise location of the Bashkirian-Moscovian boundary cannot be determined at this time.

Because *Triquitrites sculptilis* appears near the middle of an unnamed formation in the Forest City Basin of Missouri, the Bashkirian-Moscovian boundary is slightly older than the middle of the unnamed formation between the Warner Formation above and the McLouth Formation below (Peppers et al., 1993).

Appalachian coal region. The Bashkirian-Moscovian boundary is correlated with the top of the Upper Elkhorn coal bed in eastern Kentucky, the Joyner coal bed in eastern Tennessee, and the base of the No. 2 Gas coal in West Virginia. This correlation is tentative because some key taxa present in the upper Bashkirian and lower Moscovian in the Donets basin, in Illinois, and in western Europe, have not been reported in the Appalachian coal region.

The Bashkirian-Moscovian boundary is only a little older than the Westphalian B-C boundary, so that some of the same spores are useful in correlating both boundaries (Plate 2). *Triquitrites sculptilis,* which appears at about the base of the Moscovian (Teteryuk, 1976), appears in the No. 2 Gas coal bed a little below the Cedar Grove coal in West Virginia (Kosanke, 1988a). It was also recorded in the Princess No. 3 coal, which is much younger than the Cedar Grove coal. *Microreticulatisporites sulcatus* appears for the first time in the Lower Chilton coal bed, a little above the Cedar Grove coal (Eble, 1994), and in the lower Moscovian in the Donets basin.

Westphalian B-C (Duckmantian-Bolsovian) boundary

Western Europe. According to Bouroz et al. (1978), the Westphalian B-C boundary can be distinguished accurately in western Europe only where the intervening Mansfield Marine Band in Great Britain, the Aegir Marine Band in Germany, and the Rimbert Marine Band in France are present. The boundary between the stages cuts through the Middle Coal Measures in Britain, and divides the Horst and Dorsten Formations in Germany, and the Six Sillon and Pouilleuse Formations in northern France (Plate 1). The Westphalian B-C boundary is in the lower part of the *Paripteris linguaefolia* floral zone of Wagner (1984), which is transitional with the underlying floral zone. The boundary is in the upper part of the *Vestispora magna* (IX) spore assemblage zone in Britain (Smith and Butterworth, 1967) and near the top of the *Microreticulatisporites nobilis-Florinites junior* (NJ) spore assemblage zone in the scheme of Clayton et al. (1977).

Triquitrites sculptilis begins its range in late Westphalian B time in western Europe. Grebe (1972) reported a few dubious specimens of the species from the Westphalian A of the Ruhr basin. *Vestispora fenestrata, Microreticulatisporites sulcatus,*

and *Torispora securis* first appear at about the Westphalian B-C boundary, except in The Netherlands, where *V. fenestrata* appears in the middle of the Westphalian C (Van de Laar and Fermont, 1989). *Torispora securis* reaches epibole proportions in the middle of the Westphalian C. *Punctatosporites granifer* appears at a little above to a little below the Westphalian B-C boundary (Paproth et al., 1983).

Illinois basin. Palynological data support correlating the boundary between the Westphalian B and C Stages with the middle of the lower half of the Tradewater Formation, a little below the Manley Coal Member in Illinois, and a little below the Smith coal in western Kentucky and the equivalent Blue Creek Coal in Indiana (Peppers, 1984).

Shaver and Smith (1974) and Shaver (1984) correlated the Westphalian B-C boundary with the top of the Brazil Formation in Indiana, and Sohn and Jones (1984) correlated it with the top of the Abbott Formation (now the lower half of the Tradewater Formation) on the basis of studies of ostracod faunas. The boundary was placed a little above the middle Atokan by Phillips (1981) and Phillips and Peppers (1984), who used coal-ball and palynological data.

In the Illinois basin, the Westphalian B-C boundary marks the first appearance of a number of spore species followed by a rise in the abundance of fern spores (Plate 3). *Triquitrites sculptilis* begins its range below the Manley Coal in Illinois. *Vestispora fenestrata, Microreticulatisporites sulcatus,* and *Torispora securis* appear a little later than *T. sculptilis* and at about the Westphalian B-C boundary. The base of the epibole of *T. securis* is in the Tarter Coal in Illinois and in the middle Westphalian C in western Europe (Plate 2). The base of the epibole of *Punctatosporites minutus* is just below that of *Torispora. Laevigatosporites punctatus* appears in the Manley Coal in Illinois, and *Punctatosporites granifer* appears at or a little above the Westphalian B-C boundary in Great Britain (Smith and Butterworth, 1967) and Germany (Grebe, 1972). The two species are very similar in morphology and may be synonymous. Loboziak (1971) and Coquel (1976) reported *P. granifer* in older strata, but their specimens do not appear to belong to that species. *Spinosporites exiguus* (*Apiculatasporites lappites* in earlier reports by Peppers) is first observed in the Smith coal of western Kentucky, and its probable synonymous species in Europe, *Punctatosporites rotundus,* appears in early Westphalian C time (Loboziak, 1971). *Radiizonates difformis* and *R. rotatus* appear in the Mariah Hill Coal of Indiana, which is slightly older than the Tarter Coal.

Western Interior coal province. According to my interpretation, the Westphalian B-C boundary correlates with the middle of the Kilbourn Formation in Iowa and with the middle of an unnamed formation between the McLouth and Warner Sandstones in the Forest City basin in Missouri. Palynological data are not adequate to permit correlation with the southern part of the Western Interior coal province.

In the lower part of the Pennsylvanian in the Western Interior coal province, palynological correlations with the European chronology have been mainly limited to the work of Ravn (1979, 1986), who investigated the coals in the Cherokee Group of Iowa. A report (Ravn et al., 1984) on the stratigraphy and stratigraphic nomenclature of the formations and members in the Cherokee Group, which includes the Westphalian B-C boundary, was based partly on these palynological studies. Ravn (1986) indicated a higher position for the Westphalian B-C boundary than I have by correlating it with the middle of the Kalo Formation (between the Blackoak and Cliffland Coal Members) and the upper boundary of the Atokan Stage in Iowa. Ross and Ross (1988), in their compilation of worldwide transgressive-regressive sequences, incorrectly correlated the Westphalian B-C boundary with the much younger late Desmoinesian (Westphalian D) strata in the Midcontinent.

Triquitrites sculptilis, Vestispora fenestrata, and *Radiizonates difformis* first appear in unnamed coals in the Kilbourn Formation in Iowa. The apparently early occurrence of *Radiizonates difformis* at the base of the formation is explained by the fact that the sample is shale, which represents a different environment than that of coal. A hiatus must occur between the unnamed coals in the Kilbourn Formation and the Blackoak Coal, because *Torispora* and *Laevigatosporites globosus* occur for the first time and are of epibole proportions in the same coal, the Blackoak Coal. In several coals in the Kentucky part of the Illinois basin, *Torispora* and *Laevigatosporites globosus* appear in several coals below the coal in which they become abundant.

The Westphalian B-C boundary is correlated with the middle of an unnamed formation between the top of the McLouth Sandstone Formation and the base of the Warner sandstone in the Forest basin in Missouri (Peppers et al., 1993). At the middle of the formation, *Triquitrites sculptilis* also appears for the first time, and *Lycospora micropapillata* briefly becomes very abundant. This event corresponds to the epibole of the latter species near the Westphalian B-C boundary in the Illinois basin and Appalachian region (Plate 3). The stratigraphic range of *Radiizonates difformis* begins in the upper part of the unnamed formation.

Appalachian coal region. In this book, the Westphalian B-C boundary is correlated with the base of the Fire Clay coal bed in eastern Kentucky, the Windrock coal bed in eastern Tennessee, and the base of the Hernshaw coal bed in West Virginia.

The Westphalian B-C boundary is correlated partly on the basis of a comparison of the stratigraphic ranges of *Vestispora fenestrata, Triquitrites bransonii, T. sculptilis, Spinosporites exiguus,* and the epibole of *Punctatosporites minutus* in Kentucky, Tennessee (Fig. 3, Plate 3), and Europe. Gillespie and Pfefferkorn (1979) correlated, by use of plant compressions, the Westphalian B-C boundary with the Winifrede coal bed in West Virginia (Plate 1). *Vestispora fenestrata,* which appears at about the Westphalian B-C boundary elsewhere, appears for the first time in the Hernshaw coal in West Virginia (Kosanke, 1988a). *Triquitrites sculptilis,* which begins in late Westphalian B time in

Europe, was noted in the No. 2 Gas coal and then not again until the Hernshaw coal. *Laevigatosporites globosus* also appears in the Hernshaw, and the base of its epibole is in the Little Coalburg coal, just below the Coalburg coal (Kosanke, 1988a). In Europe, however, *Laevigatosporites globosus (Latosporites globosus)* appears later in the upper half of the Westphalian C (Smith and Butterworth, 1967); its epibole is not known. The base of the epibole of *Punctatosporites minutus* is in the Winifrede(?) coal (Kosanke, 1988a), which is a little above the Hernshaw coal, and in the lower part of the Westphalian C in Europe. The range of *Torispora* apparently begins later in the Appalachian coal region than in Europe and Illinois, because it first appears in the Stockton coal bed (Eble and Gillespie, 1986a; Kosanke, 1988a) and reaches epibole abundance in the Little No. 5 Block coal bed just below the Lower No. 5 Block coal. *Radiizonates difformis* first appears in the Winifrede(?) coal (Kosanke, 1988a) and in the middle Westphalian C.

In eastern Kentucky, *Vestispora fenestrata* first appears in the Little Fire Clay coal and near the Westphalian B-C boundary. Palynology of the Fire Clay coal, just above the Little Fire Clay coal and equivalent to the Hernshaw coal, was the subject of theses by Grosse (1979) and Unuigboje (1987). Grosse and Helfrich (1984) found major differences between spore assemblages above and below the Flint Clay in the Fire Clay coal. Eble et al. (1989) used palynology and coal petrography to present a palecological interpretation of the deposition of the coal. *Spinosporites exiguus* appears in the Hazard No. 4 Rider coal; *Punctatosporites rotundus* first appears near the Westphalian B-C boundary in Europe (Fig. 3). *Radiizonates difformis* extends from the Hazard No. 6 through Hazard No. 9 coal beds (Table 8) and occurs in the middle Westphalian C in Europe. Its occurrence in the Hazard No. 7 and Hazard No. 8 coals, and Princess No. 3 coal bed was documented by Helfrich (1981, 1984) and Kosanke (1981). Jennings (1981) described compression floras from the same sequence near Hazard, Kentucky, and he concluded that they are Westphalian B rather than Westphalian C in age. In eastern Tennessee, *Triquitrites bransonii* and *Vestispora fenestrata* first appear in the Upper Pioneer coal. *Laevigatosporites globosus* appears a little later, in the Windrock coal.

The relative abundance of *Lycospora pellucida, L. granulata,* and *L. micropapillata* in the Illinois Basin, eastern Kentucky, and eastern Tennessee (Plate 3) is also an aid in correlating coals between the two regions. *Lycospora pellucida* was generally dominant in the Illinois basin and Kentucky in the lower part of the Pennsylvanian up to about midway between the Bell and Smith coals (middle Westphalian B), where *L. granulata* became most abundant. *Lycospora micropapillata* became very abundant from a little below to a little above the Smith coal, and then *L. granulata* again became dominant until the end of the Desmoinesian and Westphalian. Kosanke (1988a) reported *L. micropapillata* as being abundant for the first time in the No. 2 Gas coal (late Westphalian B) in West Virginia. It remained abundant up to the Stockton "A" coal bed, which is just above the Stockton coal and is late Westphalian C in age.

Donets basin. The Westphalian B-C boundary is correlated with Limestone K_7 near the top of the C_2^5 suite and just above the base of the Moscovian Stage.

Teteryuk (1976) correlated the Westphalian B-C boundary with Limestone K_1 in the upper part of the Bashkirian in the Donets basin; he later (1982) correlated it with Limestone K_3, which marks the Bashkirian-Moscovian boundary. Inosova and others (1975, 1976) did not subdivide the Westphalian into stages, but they showed that *Vestispora fenestrata* and *Torispora* began their ranges at about Limestone L_3, which would correspond to the lower part of the Westphalian C in western Europe. *Punctatosporites minutus* also reached epibole proportions at Limestone L_3 and in the lower part of the Westphalian C in Europe.

Atokan-Desmoinesian Stage boundary

Western Interior coal province. Identification of the boundary between the Atokan and Desmoinesian Stages in the Midcontinent has been a problem because of the previously mentioned uncertainties regarding the stratigraphic limits of the Atokan Stage and because no specific type section, other than exposures along the Des Moines River in Iowa, was designated for the Desmoinesian Series (Keyes, 1893). The Desmoinesian was to include all rocks down to the base of the Pennsylvanian in Iowa; thus, the Desmoinesian included rocks of Morrowan and Atokan age (Ravn et al., 1984; Ravn, 1986).

Determining the position of the Atokan-Desmoinesian boundary varies according to whether one accepts the original definition or a boundary derived from paleontological criteria. In proposing the Atokan Series, Spivey and Roberts (1946) defined its upper boundary as the base of the Hartshorne Sandstone in the Arkoma basin in south-central Oklahoma. Because of inadequate palynological data from the upper part of the Atoka Formation, direct palynological correlation of the boundary with strata in other regions is difficult. The palynological correlations that do exist, however, reinforce correlations determined from fusulind and conodont studies.

Sutherland and Manger (1984) pointed out that stratigraphers and paleontologists in the western part of the Midcontinent recognize the base of the zone of *Fusulina (Beedeina)* and *Wedekindellina* as the base of the Desmoinesian. However, the "Griley" Limestone near the top of the Atoka Formation in its type area contains *Fusulina (Beedeina)* and advanced forms of *Fusulinella*, indicating an overlap of the Atokan and Desmoinesian (Douglass and Nestell, 1984; Zachry and Sutherland, 1984).

The upper part of the Atokan is in floral zone 8 (*Neuropteris tenuifolia*) of Read and Mamay (1964); the lower part of the Desmoinesian is in floral zone 9 (*Neuropteris rarinervis*).

Wilson (1976b), who studied spores macerated from clastic rocks in the Atoka Formation at its type area, thought the presence of *Zosterosporites triangularis* indicated the Desmoinesian

character of the spore assemblage because the species had been described from the seat rock of the Princess No. 5B coal bed of Kentucky (Kosanke, 1973). Kosanke correlated the latter coal with the Davis Coal Member of Illinois, which is a little below the middle of the Desmoinesian. These specimens, however, could have been redeposited from older strata. Ravn (1986) reported that the species ranges from the Morrowan Wildcat Den Coal Member to the early Desmoinesian Laddsdale Coal Member in Iowa. *Zosterosporites triangularis* extends from the Pope Creek Coal (upper Atokan) to the Hermon Coal Member (lower Desmoinesian) in Illinois.

Wilson (1976a) summarized the palynology of Desmoinesian coals in northeastern Oklahoma from numerous palynological reports on individual coal beds. The distribution of spores in those coals is similar to that in coals in the upper part of the Tradewater Formation of Illinois. Spore assemblages in the Lower and Upper Hartshorne coals at the base of the Desmoinesian Stage in Oklahoma reported by Wilson (1970) are not sufficiently diagnostic, however, to permit precise correlation with coal beds in Illinois. I studied the palynology of a sample of the Hartshorne coal from an outcrop in Harris County, Oklahoma (Table 9), and it indicates that the Hartshorne coal is correlative with coals in Illinois that are just above the Seville Limestone Member.

A sample from a coal just below the Warner Sandstone Member in the McAlester Formation, which overlies the Hartshorne Formation, was obtained from a diamond-drill core from Muskogee County, Oklahoma. Only one specimen of *Endosporites zonalis* was observed in maceration 2725A (Table 9), but its presence indicates that the coal is older than the Lewisport coal in the Illinois basin. *Murospora kosankei*, which also occurs in maceration 2725A, first appears in the Hermon Coal just above the Seville Limestone in Illinois, which is regarded as the top of the Atokan.

In Oklahoma (Fay et al., 1979), the Riverton coal has been considered to be younger than the Hartshorne Formation and older than the McAlester Formation. Rosowitz (1982) correlated the Riverton coal in southeastern Kansas with the Lower Hartshorne coal of Oklahoma because of the percentage of similar spore genera. The presence of *E. zonalis* (illustrated as *E. globiformis*) and scarcity of *Thymospora pseudothiessenii* in the Riverton supports the conclusion that it is as old as or almost as old as the Hartshorne coal.

Searight and Howe (1961) regarded the Riverton Formation in Missouri as underlying the Hartshorne Formation because T. L. Thompson (unpublished data) had found Atokan conodonts at two localities containing shale, later identified as Riverton shale (Thompson, 1979). Lambert and Thompson (1990) confirmed the presence of a late Atokan fauna from the lower part of what is called the Riverton Formation in southwestern Missouri. Marine fossils have not been found in the type area of the Riverton Formation in southeastern Kansas. According to Nuelle (Missouri Geological Survey, 1989, personal communi-

TABLE 8. SPORE ANALYSIS OF HAZARD COAL BEDS IN HAZARD AND PERRY COUNTIES, KENTUCKY

Maceration Spore taxa*	Hazard No. 4			Hazard No. 4 Rider 2666		Hazard No. 6		
	E	F	G	H	J	K	L	M
Deltoidospora adnata			X†					
D. priddyi	0.5	X	2.0	0.5		X	0.5	1.0
D. pseudolevis								
D. sphaerotriangula			X	0.5	0.5		0.5	
D. subadnatoides	2.5							
D. spp.	1.0				0.5	1.5		
Punctatisporites cf. *edgarensis*					X			
P. flavus								
P. glaber		X						
P. minutus						1.0		
P. obesus	X		X	X	X			
P. spp.								0.5
Calamospora breviradiata			X			0.5		2.0
C. hartungiana	X				X		X	
C. liquida								X
C. mutabilis								
C. straminea				X				0.5
Granulatisporites adnatoides	0.5				0.5			
G. granularis	0.5	X	X		X	X		
G. granulatus					0.5	X		0.5
G. micrograniter	1.5			1.0	0.5			
G. minutus	1.5	0.5		0.5	1.5		0.5	
G. pallidus	X				X	X	1.5	0.5
G. verrucosus							0.5	
G. spp.				0.5		1.0	1.5	
Cyclogranisporites aureus		X	X	X	X	0.5	X	0.5
C. microgranus								
C. minutus								
C. obliquus								
Sinuspores sinuatus		X	X		X			
Verrucosisporites microtuberosus	X	1.0	2.0		1.0	X	0.5	
V. sifati							0.5	
V. verrucosus		X						
V. cf. *verrucosus*		X	X			X		
Kewaneesporites patulus	X							
Lophotriletes commissuralis	0.5					0.5	0.5	
L. microsaetosus			0.5		0.5			
L. mosaicus					X			
L. spp.	0.5					0.5		
Anapiculatisporites spinosus	0.5							
Apiculatasporites latigranifer							X	
A. spp.	0.5					0.5	X	
Planisporites granifer	X			X				
Pilosisporites aculeolatus	2.0	2.0	1.0		1.5	X	0.5	0.5
P. williamsii					X	1.0		
P. spp.	0.5		0.5	1.0	2.5			
Raistrickia breveminens	0.5			X		0.5	X	
R. crocea	X				0.5			
R. superba								
Spackmanites habibii			0.5			0.5		
Convolutispora florida								
C. spp.	X							

TABLE 8. SPORE ANALYSIS OF HAZARD COAL BEDS IN HAZARD AND PERRY COUNTIES, KENTUCKY (continued - page 2)

Maceration Spore taxa*	Hazard No. 7 2666			Hazard No. 8 2671				Hazard No.9 2672		
	N	O	P	A	B	C	D	A	B	C
Deltoidospora adnata										
D. priddyi	0.5		0.5	0.5	1.0		0.5	X†	0.5	
D. pseudolevis					X					
D. sphaerotriangula				X						
D. subadnatoides	0.5		1.0					1.0		0.5
D. spp.				1.0		1.5				
Punctatisporites cf. *edgarensis*										
P. flavus		X								
P. glaber										
P. minutus		0.5				2.0		1.0		
P. obesus				X	0.5	X	X	X		
P. spp.			X							
Calamospora breviradiata				0.5	2.0	0.5	1.0			
C. hartungiana				X	0.5		0.5			
C. liquida										
C. mutabilis				X	X					
C. straminea	X									
Granulatisporites adnatoides									0.5	
G. granularis				0.5						
G. granulatus										0.5
G. micrograniter										1.0
G. minutus	0.5	1.5	1.0	0.5				1.0	0.5	0.5
G. pallidus			X	1.0						
G. verrucosus										
G. spp.	0.5			1.0	1.0		0.5	0.5	0.5	
Cyclogranisporites aureus				X			X	X		
C. microgranus						X				
C. minutus					0.5					
C. obliquus	X							0.5		
Sinuspores sinuatus										
Verrucosisporites *microtuberosus*		X	X		X		X	0.5	4.0	
V. sifati								X	0.5	
V. verrucosus									0.5	
V. cf. *verrucosus*										
Kewaneesporites patulus										
Lophotriletes commissuralis		0.5								
L. microsaetosus		1.0		1.0		0.5		0.5		
L. mosaicus										
L. spp.	0.5									
Anapiculatisporites spinosus							0.5			
Apiculatasporites latigranifer										
A. spp.										
Planisporites granifer		X								
Pilosisporites aculeolatus	0.5		1.0							
P. williamsii	X	0.5							0.5	
P. spp.	0.5	0.5					0.5		1.0	
Raistrickia breveminens	X									
R. crocea					X	X				
R. superba							X			
Spackmanites habibii										
Convolutispora florida	X									
C. spp.										

TABLE 8. SPORE ANALYSIS OF HAZARD COAL BEDS IN HAZARD AND PERRY COUNTIES, KENTUCKY (continued - page 3)

Maceration Spore taxa*	Hazard No. 4			Hazard No. 4 Rider 2666	Hazard No. 6			
	E	F	G	H	J	K	L	M
Microreticulatisporites nobilis								
Dictyotriletes bireticulatus						0.5	0.5	
Camptotriletes bucculentus								
Ahrensisporites guerickei	X							
A. guerickei var. ornatus	X		X					X
Triquitrites additus								
T. bransonii			X	0.5		X	X	
T. exiguus								
T. sculptilis						X	0.5	0.5
Zosterosporites triangularis		X		X				
Reinschospora magnifica						X		
Knoxisporites triradiatus			X	X	X	X	X	X
Reticulatisporites mediareticulatus	0.5	0.5						
R. muricatus						X		X
R. polygonalis				X		X		
R. reticulatus			X			0.5	X	X
R. reticulocingulum		X					X	
Reticulitriletes falsus			X					
R. spp.			1.0					
Savitrisporites concavus	2.5			1.0				
S. nux	0.5							X
Grumosisporites varioreticulatus								
Crassispora kosankei								
Granasporites medius			0.5					
Simozonotriletes intortus				X				
Densosporites annulatus	4.5	X		25.0	2.0		15.0	X
D. sphaerotriangularis	1.5		6.0		1.0	0.5	2.0	
D. triangularis			4.0					
D. spp.		0.5	0.5	0.5				
Lycospora granulata	23.0	65.0	12.0	45.0	27.5	36.5	24.5	49.0
L. micropapillata	7.5	3.0	0.5		2.0		2.0	14.5
L. noctuina		X						
L. pellucida	1.0	9.0	3.0	2.5	2.5	2.0	5.0	14.5
L. pusilla	0.5	9.0	1.5	7.5	1.5	1.0	1.5	
L. rotunda			0.5		3.5	1.0		0.5
L. subjuga		1.0	2.0		0.5			
L. spp.					0.5			
Cristatisporites indignabundus	3.5		47.5		1.5	0.5		
Cirratriradites annulatus								
C. annuliformis								X
C. maculatus	X		X					
C. saturnii	X	X	X		X	X		
Radiizonates difformis							1.5	
Endosporites globiformis	1.0	X	0.5	X			0.5	2.5
E. plicatus		X	X		0.5	0.5		
E. zonalis					X	X		
Alatisporites hexalatus								
A. hoffmeisterii								
A. pustulatus						X	X	
A. trialatus								X
Laevigatosporites desmoinesensis	3.0	1.0	4.0		7.0	1.5	3.5	
L. globosus					8.0	15.5	3.0	

TABLE 8. SPORE ANALYSIS OF HAZARD COAL BEDS IN HAZARD AND PERRY COUNTIES, KENTUCKY (continued - page 4)

Maceration Spore taxa*	Hazard No. 7 2666			Hazard No. 8 2671				Hazard No.9 2672		
	N	O	P	A	B	C	D	A	B	C
Microreticulatisporites nobilis							X			
Dictyotriletes bireticulatus					0.5					
Camptotriletes bucculentus									X	0.5
Ahrensisporites guerickei										
A. guerickei var. *ornatus*		X								
Triquitrites additus	X									
T. bransonii	X									
T. exiguus						X				
T. sculptilis		0.5	X	X	X	0.5	1.0	0.5	0.5	
Zosterosporites triangularis										
Reinschospora magnifica										
Knoxisporites triradiatus	X									
Reticulatisporites mediareticulatus										
R. muricatus										
R. polygonalis										
R. reticulatus										
R. reticulocingulum										
Reticulitriletes falsus										
R. spp.										
Savitrisporites concavus										
S. nux										
Grumosisporites varioreticulatus	X									
Crassispora kosankei	0.5			1.0	2.5	X				
Granasporites medius	1.0				1.0		0.5			
Simozonotriletes intortus										
Densosporites annulatus	12.0		0.5	X	X				0.5	
D. sphaerotriangularis	0.5		5.5							0.5
D. triangularis	1.0	0.5								
D. spp.	0.5								1.5	1.0
Lycospora granulata	42.0	11.0	4.5	37.0	27.0	28.5	70.0	12.0	9.5	14.5
L. micropapillata	2.5			12.5	13.5	58.0	6.0	0.5		
L. noctuina										
L. pellucida			0.5	29.5	29.0	3.5	7.0	1.5		2.5
L. pusilla		1.0	0.5	0.5	5.0	2.0	1.5		1.0	0.5
L. rotunda										
L. subjuga										
L. spp.										
Cristatisporites indignabundus			1.0							
Cirratriradites annulatus									X	
C. annuliformis										
C. maculatus										
C. saturnii				X	X		X			X
Radiizonates difformis		1.0	16.5					31.0	11.0	
Endosporites globiformis	X		X	0.5	1.0	0.5				0.5
E. plicatus	X		X							
E. zonalis		X								
Alatisporites hexalatus									X	
A. hoffmeisterii								0.5		
A. pustulatus										
A. trialatus										
Laevigatosporites desmoinesensis		1.5		1.5				X	X	1.5
L. globosus	22.5	67.0	56.5	1.0	X		3.0	37.0	50.0	72.5

TABLE 8. SPORE ANALYSIS OF HAZARD COAL BEDS IN HAZARD AND PERRY COUNTIES, KENTUCKY (continued - page 5)

Maceration Spore taxa*	Hazard No. 4			Hazard No. 4 Rider 2666		Hazard No. 6		
	E	F	G	H	J	K	L	M
L. medius	0.5	1.0	2.0	1.0		2.5	0.5	0.5
L. ovalis	19.5	5.0	7.0	7.5	19.5	15.5	26.0	7.5
L. striatus				X		X		X
L. vulgaris	0.5			X		X		
Punctatosporites minutus	8.5	1.5		4.0	9.5	13.5	4.5	4.0
Spinosporites exiguus				0.5				
Torispora securis								
Dictyomonolites swadei					0.5			
Vestispora clara					X			
V. fenestrata							X	
V. pseudoreticulata							1.0	X
V. sp.								
Florinites mediapudens		X		2.0	1.5		1.5	1.0
F. visendus			X					X
F. volans	6.0			X				

cation), the shale called the Riverton shale in Missouri may correlate with the McAlester Formation in Oklahoma, which overlies the Hartshorne Formation.

Pierce and Courtier (1937) named the Riverton coal for a 10-in-thick (25.4 cm) coal occurring in a sink hole in Cherokee County, Kansas. Maceration 152 of this coal contains an early Desmoinesian spore assemblage (Table 10). Two coal seams in the upper part of the Riverton Formation were also sampled in a mine in Barton County, Missouri. The 1 ft (0.31 m) coal at the top of the formation contains a poorly preserved spore assemblage. The 15 in (38 cm) coal (maceration 3118A) 12 ft (3.66 m) below the upper one contains a well-preserved early Desmoinesian spore assemblage (Table 10). As many as three coal seams occur in the upper part of the Riverton Formation in Missouri (Searight and Howe, 1961).

I also examined another sample of the Riverton coal in Barton County, Missouri, and *Lycospora* constitutes more than three-fourths of the spore assemblage in the lower bench (maceration 2968, Table 10). The upper bench (maceration 2967) has an unusually large amount of *Endosporites globiformis*, and *Laevigatosporites globosus* is also abundant. *Dictyotriletes bireticulatus*, which ranges up to the Rock Island Coal, was not recorded. The complete spore assemblage indicates that the Riverton coal of Missouri is Desmoinesian in age.

In the Forest City basin in Missouri, the Atokan-Desmoinesian boundary is just below the base of the Warner sandstone. In outcrops in Kansas, Oklahoma, and Missouri, palynological evidence indicates that the Riverton coal is Desmoinesian in age, but other interpretations consider the Riverton Formation to be Atokan in age. A coal just below the Warner sandstone in a core drilled in the Forest City basin (Peppers et al., 1993) may correlate with the Riverton coal of Kansas. The

Missouri Geological Survey, which is revising the stratigraphy of the entire Pennsylvanian in its state, is considering this problem. *Radiizonates difformis, Densoporites annulatus,* and *Dictyotriletes bireticulatus* extend to the top of the Riverton Formation as currently defined by the Missouri Geological Survey. *Murospora kosankei, Vesicaspora wilsonii,* and *Vestispora clara* also appear in the coal just above the Warner sandstone of Missouri. These species occur for the first time at or near the base of the Desmoinesian in the Illinois basin.

Morgan (1955a, 1955b) studied the palynology of the lower Desmoinesian McAlester, Stigler, Riverton, and Rowe coals in Oklahoma and correlated the McAlester with the Stigler: however, he did not make correlations for outside of Oklahoma. His limited amount of data supports a correlation of the McAlester coal with about the Lewisport (Dawson Springs) Coal in the Illinois basin. Dempsey (1964, 1967), who compared spore assemblages in the McAlester coal with those in several coals older than the Seville Limestone in Illinois, concluded that the McAlester coal does not correlate with any of those coals. Maceration 3153A (Table 11) of the lower 4 in (10 cm) of the McAlester coal from Pittsburg County, Oklahoma, contains *Thymospora pseudothiessenii*, which first appears just above the Seville Limestone in Illinois. *Vestispora wanlessii* was observed in maceration 3153B of the upper 3 in (7.6 cm) of coal, and it extends from the Hermon Coal just above the Seville Limestone to the New Burnside Coal Bed in Illinois. The lack of *Endosporites zonalis* and the low abundance of *Laevigatosporites globosus* and *Torispora securis* also indicate that the McAlester coal is at least as young as the Lewisport coal bed in the Illinois basin.

A coal in the northern part of the Ardmore basin was formerly thought to be in the Atokan Bostwick Member of the Lake Murray Formation, but was more recently called Unnamed

TABLE 8. SPORE ANALYSIS OF HAZARD COAL BEDS IN HAZARD AND PERRY COUNTIES, KENTUCKY (continued - page 6)

| Maceration | Hazard No. 7 | | | Hazard No. 8 | | | | Hazard No.9 | | |
| | 2666 | | | 2671 | | | | 2672 | | |
Spore taxa*	N	O	P	A	B	C	D	A	B	C
L. medius								1.5	0.5	
L. ovalis	5.0	5.0	10.0	8.5	7.0	1.0	2.5	4.0	7.0	2.5
L. striatus										
L. vulgaris					X					
Punctatosporites minutus	9.0	7.0		1.0	3.0	1.0	3.5	4.0	8.0	
Spinosporites exiguus					0.5			0.5		
Torispora securis									X	
Dictyomonolites swadei										
Vestispora clara										
V. fenestrata				X		X				
V. pseudoreticulata			X	X	0.5	X	X			
V. sp.								X		
Florinites mediapudens			0.5	1.5	4.0	0.5	2.0		X	
F. visendus										
F. volans										

*Formal systematics of taxa given in Appendix A.
†X = present but not in count.

Unit 3 (Sutherland and Grayson, 1992). This coal was dated as Desmoinesian by Rashid (1968) and Wilson and Rashid (1982) because of the presence of some species of spores cited as having been previously reported only from the Desmoinesian or younger rocks. Some of the species are now also known to occur in Atokan rocks (Ravn, 1979, 1986; Peppers, 1984, and unpublished data), but one of the species, *Cadiospora*, is a reliable indicator of a Desmoinesian or younger Pennsylvanian age (Plate 2). Rashid also stated that there was an associated plant compression assemblage similar to the flora of the *Neuropteris rarinervis* zone of Read and Mamay (1964), indicating a Desmoinesian age. This flora also occurs in the Hartshorne Formation (Hendricks and Read, 1934).

A coal in Carter County, Oklahoma, at section 363 (Tennant, 1981) and stop 7 (Tennant et al., 1982), about 14 mi (22.4 km) northeast of the section that Rashid (1968) studied, was also thought to be in the Bostwick Member. The bottom half of the coal (maceration 3035B) is interbedded carbonaceous shale and coal. Its spore assemblage is more diverse and is composed of more fern spores than the assemblage in the upper portion (maceration 3035C), which is bright-banded coal (Table 12). Spore assemblages reveal that the coal is correlative with the interval between the Lewisport and New Burnside Coals in Illinois. *Vestispora wanlessii*, which also occurs in the McAlester coal (Table 11), was noted in the "Bostwick" coal and extends from about the Hermon to the New Burnside Coal in Illinois. *Lophotriletes ibrahimii* and *Murospora kosankei* also occur in maceration 3035B and first appear in the lower Desmoinesian in Illinois. Only one specimen of *Thymospora pseudothiessenii* was observed in the coal, and it is rare in the McAlester coal and other

lower Desmoinesian coals. *Dictyotriletes bireticulatus* and *Endosporites zonalis* are absent in the "Unnamed Unit 3" and McAlester coals but extend from the Lower Pennsylvanian to just above the base of the Desmoinesian in the Illinois basin. *Densosporites annulatus* and *Radiizonates*, which range up to the late Atokan, are also absent.

The coal samples that Rashid and I studied are probably slightly younger than the Hartshorne coal (Table 9) and are early Desmoinesian in age. From lithostratigraphic and biostratigraphic studies, Clopine (1986, 1991) and Sutherland (1990) concluded that the Bostwick Member is cut out in the northern part of the Ardmore basin.

The lower part of the Cherokee Group in Iowa (Ravn et al., 1984; Ravn, 1986) and Kansas (Wilson, 1979c) was originally designated as early Desmoinesian (Keyes, 1893). It contains spore assemblages that are Atokan in age and equivalent to the middle of the Tradewater Formation in Illinois. The Des Moines Supergroup was erected by Ravn et al. (1984) to include the Atokan and Desmoinesian Series in Iowa. Shaver (1984) felt that there is little need for the Atokan because the Desmoinesian was originally defined to include strata as old as the base of the Pennsylvanian in Iowa. However, Shaver et al. (1985), working on the COSUNA correlation charts, as well as geologists of the Iowa Geological Survey (Ravn, 1979, 1981, 1986), and Ravn et al. (1984) proposed a compromise. They located the Atokan-Desmoinesian boundary between the Blackoak and Cliffland Coals on the basis of the last occurrence of *Dictyotriletes bireticulatus*, which is in the Blackoak Coal in Iowa. They did not specify, however, which unit marks the boundary between the two coals. Ravn (1979) performed a detailed study of the

R. A. Peppers

TABLE 9. SPORE ANALYSIS OF THE HARTSHORNE COAL BED IN HARRIS COUNTY, OKLAHOMA (MACERATION 3121B), AND OF A COAL (MACERATION 2725A) BELOW THE WARNER SANDSTONE IN MUSKOGEE COUNTY, OKLAHOMA

Spore taxa*	Macerations		Spore taxa*	Macerations	
	3121B[†] (%)	2725A (%)		3121B[†] (%)	2725A (%)
Deltoidospora sphaerotriangula	1.0		Reticulatisporites reticulatus		X
Punctatisporites flavus	1.0		Crassispora kosankei	X	
P. glaber	0.5		Granasporites medius		1.5
P. minutus	1.5		Murospora kosankei		X
Calamospora breviradiata	1.5		Densosporites sphaerotriangularis		1.0
C. hartungiana	1.0				
			Lycospora granulata	27.0	44.0
Granulatisporites granularis	0.5	0.5	L. pellucida	2.0	0.5
G. micrograni fer	X[§]	X	L. pusilla	X	0.5
G. minutus		X	L. rotunda	2.0	3.0
G. verrucosus		X			
			Cirratriradites annulatus	X	
Cyclogranisporites leopoldi		0.5	C. annuliformis		X
C. microgranus		X	C. saturnii	0.5	
C. minutus	2.0		Endosporites globiformis	1.0	
C. obliquus	3.5		E. zonalis		X
C. orbicularis	2.5		Alatisporites pustulatus		X
Verrucosisporites microtuberosus	X		Laevigatosporites desmoinesensis	18.5	
V. cf. verrucosus		0.5	L. globosus	12.0	33.5
Lophotriletes commissuralis	X		L. ovalis	9.5	1.5
L. microsaetosus	2.0	0.5	L. vulgaris	X	
L. pseudaculeatus	X		Latosporites minutus	1.0	
			Punctatosporites minutus	3.5	8.0
Anapiculatisporites edgarensis	0.5		Spinosporites exiguus	0.5	
A. spinosus		X	Torispora securis	1.0	X
Apiculatasporites setulosus	X		Vestispora fenestrata	X	0.5
Raistrickia breveminens	0.5				
R. crocea	X		Florinites mediapudens	0.5	3.5
			F. similis		0.5
Microreticulatisporites nobilis		X	F. volans	0.5	X
M. sulcatus	X	X	F. sp.	0.5	
Triquitrites bransonii	2.0	X	Wilsonites vesicatus		X
T. tribullatus	X	X			
Zosterosporites triangularis	X				

*Formal systematics of taxa given in Appendix A.
[†]Spores in maceration 3121A from shale underlying Hartshorne coal are poorly preserved.
[§]X = present but not in count.

palynology of the Blackoak Coal, which correlates approximately with the Pope Creek Coal of Illinois. The Cliffland Coal, which overlies the Blackoak, was correlated by Ravn (1986) with the Rock Island Coal, partly because of their relative stratigraphic positions. The Cliffland Coal may be a little younger, however, than the Rock Island Coal. The ISGS placed the Atokan-Desmoinesian boundary at the top of the Murray Bluff Sandstone, which lies between the Pope Creek and Rock Island Coals (Willman et al., 1967). Thus, the Iowa Geological Survey adopted a position for the boundary that is about the same age as that adopted by the ISGS.

Ravn (1981) placed the Atokan-Desmoinesian boundary in Iowa at the last occurrence of *Dictyotriletes bireticulatus*, which is at the top of the Blackoak Coal. In Illinois *D. bireticulatus* last occurs in the Rock Island Coal just below the Seville Limestone. In Indiana it extends as high as the "Indiana Coal II" (informal name, Shaver et al., 1986), which lies just above the Perth Limestone Member and is correlative with the Seville Limestone in Illinois. Thus, the stratigraphic range of *D. bireticulatus* extends slightly higher from west to east. Shaver (1984) indicated in an illustration that *D. bireticulatus* ranges up through the Blackoak Coal, through an unnamed overlying limestone, and to the base of the Cliffland Coal, which places the Atokan-Desmoinesian boundary in Indiana in almost the same position as that adopted by the ISGS. The first appearance of *Murospora kosankei* and possible first appearance of *Mooreisporites inusitatus* in the Cliffland Coal and the maximum abundance of *Torispora* through the Cliffland Coal (Ravn, 1986) further indicate that the coal is a little younger than the Rock Island Coal. Therefore, I am correlating the boundary in Iowa with the base of the Cliffland Coal.

Lambert (1989) proposed placing the Atokan-Desmoinesian boundary at the limestone overlying the Cliffland Coal on the basis of the ranges of species of the conodonts *Neognathodus* and *Idiognathodus*. In 1990 and 1992 Lambert placed the boundary in a slightly different position, at the base of the Cliffland Coal, and closer to the position proposed by the Iowa Geological Survey.

Illinois basin. On the basis of palynological and fusulinid evidence, I am correlating the boundary between the Atokan and Desmoinesian Stages of the Midcontinent region with the tops of the Seville Limestone Member in Illinois, the Perth Limestone in Indiana, and the Curlew Limestone Member in western Kentucky. The boundary is also correlated with the Westphalian B-C boundary, which is discussed in the following section. Correlation of the Atokan-Desmoinesian boundary in the Western Interior coal province in relation to the Illinois basin was partly discussed in the preceding section.

The Atokan-Desmoinesian boundary as adopted by the ISGS (Willman et al., 1967; Hopkins and Simon, 1975; Atherton and Palmer, 1979; Peppers and Popp, 1979) has been correlated with what was formerly the boundary between the Abbott and Spoon Formations at the top of the Murray Bluff Sandstone Member in southern Illinois and the Bernadotte Sandstone Member in northwestern Illinois. These sandstones underlie the Rock Island Coal and Seville Limestone. Shaver (1984) and Shaver et al. (1985) correlated the boundary with the top of the Minshall Coal Member, which correlates with the Rock Island Coal.

Proposed positions of the Atokan-Desmoinesian boundary in the Illinois basin have varied from the top of the Caseyville Formation (Wanless, 1956; Kosanke et al., 1960) to the top of the Seville Limestone (Douglass, 1979; Rice et al., 1979b; Williams et al., 1982).

Many stratigraphers working in the Illinois basin place the base of the zone of *Fusulina (Beedeina)* in the lower part of the Desmoinesian. Thompson (1948) considered the Atokan as the zone of *Profusulinella* and *Fusulinella*. *Fusulinella*, however, is reported as extending from the lower Atokan to the lower Desmoinesian (Douglass and Nestell, 1984). Thompson and Riggs (1959) considered the Curlew and Seville Limestones in the Illinois basin to be Desmoinesian in age because of the presence of *Fusulinella iowensis* in the limestones (Thompson, 1934). This position was supported by Thompson et al. (1959), Shaver and Smith (1974), Shaver (1984), and Shaver et al. (1985, 1986). However, *F. iowensis* was described by Thompson (1934) from lower Cherokee rocks of Iowa, which were included in the Desmoinesian Series as originally described (Keyes, 1893), but which are also known to be late Atokan in age. The Seville and Curlew Limestones were assigned to the Atokan by Cooper (1946), on the basis of ostracod studies, and by Douglass (1979, 1987) and Douglass and Nestell (1984) from fusulinid studies. Douglass (1979) noted that the fusulinid faunas, including *Fusulinella iowensis, F. iowensis stouti,* and *Millerella* in the Curlew Limestone as well as equivalent limestones in Illinois and Indiana, are the same age as those in the oldest fusulinid-bearing

limestone in Iowa and the upper part of the "Derryan" described by Thompson (1942, 1948). More recently, a *Fusulinella* fauna from near the top of the Atoka Formation in the northeastern part of the Arbuckle Mountains in Oklahoma was correlated with the Curlew Limestone fauna from its type section in western Kentucky (Douglass and Nestell, 1984). The Mitchellsville Limestone Bed (Nelson and Lumm, 1990; Nelson et al., 1991), formerly the "Curlew Limestone" of southern Illinois, is younger than the type Curlew and contains *Beedeina* (Douglass, 1979; Douglass and Nestell, 1984). Thus, according to fusulinid evidence, the Atokan-Desmoinesian boundary in the Illinois basin could be as low as the top of the Curlew or Seville Limestone or as high as the base of the Mitchellsville Limestone, which overlies the Delwood Coal Bed.

No coal occurs just below the Curlew Limestone at its type outcrop in western Kentucky. The spore assemblage in a coal (Table 13) below the Curlew Limestone and less than 1 mi (1.6 km) from the Curlew type section indicates that the coal is equivalent to the Rock Island Coal, which underlies the Seville Limestone in northwestern Illinois. This finding confirms that the Curlew Limestone correlates with the Seville and Perth Limestones. The coal is at a depth of 392.68 to 394.31 ft (119.7 to 120.2 m) in the Gil 15 core, which was described by Williams et al. (1982). In their correlation chart, Peppers and Popp (1979) erroneously showed that the Curlew Limestone is younger than the Seville and Perth Limestones. Unpublished palynological studies had indicated (Peppers) that the Lewisport (Dawson Springs) coal bed is younger than the Rock Island and Minshall Coals. Additional investigation (Peppers, 1995) confirms this stratigraphic relation. Because the Kentucky Geological Survey (Williams et al., 1982; Greb et al., 1992) and the USGS (Kehn et al., 1967; Rice et al., 1979a) accepted that the Dawson Springs coal is overlain by the Curlew Limestone, Peppers and Popp (1979) assumed that the Curlew is also younger than the Seville and Perth Limestone, which overlies the Rock Island and Minshall Coals, respectively. Douglass (1987) found, according to fusulinid evidence, that the limestone overlying the Lewisport coal is younger than the Curlew Limestone. The Lewisport coal is also younger than the coal that was formerly called "Indiana Coal II" (unofficial name, Shaver et al., 1970), which directly overlies the Perth Limestone. The "Curlew Limestone" of southeastern Illinois, which is younger than the type Curlew Limestone, was correlated by Douglass (1987), on the basis of fusulinid content, with the Putnam Hill limestone in the basal part of the Allegheny Group in Ohio. Wanless (1975) suggested that the younger "Curlew Limestone" of southern Illinois (now called Mitchellsville Limestone) is correlative with the Creal Springs Limestone Member in southeastern Illinois but, as indicated by Peppers and Popp (1979), the Mitchellsville Limestone is actually intermediate in age between the Creal Springs Limestone and type Curlew Limestone.

The name "Dawson Springs Coal" was used by Peppers (1984) for the first time in Illinois to designate a coal in south-

TABLE 10. SPORE ANALYSIS OF THE RIVERTON COAL FROM ITS TYPE SECTION IN CHEROKEE COUNTY, KANSAS (MACERATION 152), UPPER BENCH (MACERATION 2967), AND LOWER BENCH (MACERATION 2968) OF THE RIVERTON COAL BED, LOWER BENCH (MACERATION 3118A) OF THE RIVERTON COAL IN BARTON COUNTY, MISSOURI, AND OF THE DRYWOOD COAL BED (MACERATION 902) IN BARTON COUNTY, MISSOURI

Spore taxa*	Macerations				
	152[†] (%)	2967 (%)	2968 (%)	3118A (%)	902 (%)
Deltiodospora levis				X[§]	X
D. priddyi				0.5	
D. sp.				0.5	
Punctatisporites cf. edgarensis		X			X
P. flavus	X				0.5
P. glaber	X			3.0	
P. minutus		2.0	X	4.5	1.0
P. obesus	X				X
Calamospora breviradiata	X		X		0.5
C. hartungiana	X	X			X
C. pedata					X
C. straminea				X	
Granulatisporites parvus	X				X
G. minutus		X			
G. pallidus			X		0.5
G. verrucosus	X			X	
G. sp.	X				
Cyclogranisporites aureus			1.0		
C. microgranus					X
C. orbicularis				0.5	0.5
C. staplinii				0.5	0.5
Verrusosisporites microtuberosus	X		X		X
V. sifati			X	X	0.5
V. cf. verrucosus				1.0	0.5
V. sp.				0.5	
Lophotriletes commissuralis		X	2.0		2.5
L. microsaetosus	X	0.5	1.0		
Anapiculatisporites grundensis			X		X
Apiculatasporites setulosus					X
Pilosisporites aculeolatus		X	X		
Raistrickia abdita			X	X	
R. aculeolata				X	
R. breveminens			X		
R. crocea	X		X	X	0.5
R. subcrinita				X	
Microreticulatisporites lunatus					X
M. nobilis		X	X	X	1.5
M. sulcatus					3.0
Camptotriletes confertus					2.0
Ahrensisporites guerickei					X
Triquitrites additus			X		X
T. bransonii		X	1.0	X	3.0
T. exiguus					1.5
T. pulvinatus				X	
T. sculptilis	X			0.5	1.5

**TABLE 10. SPORE ANALYSIS OF THE RIVERTON COAL FROM ITS TYPE SECTION IN
CHEROKEE COUNTY, KANSAS (MACERATION 152), UPPER BENCH (MACERATION 2967),
AND LOWER BENCH (MACERATION 2968) OF THE RIVERTON COAL BED, LOWER BENCH
(MACERATION 3118A) OF THE RIVERTON COAL IN BARTON COUNTY, MISSOURI,
AND OF THE DRYWOOD COAL BED (MACERATION 902) IN BARTON COUNTY, MISSOURI**
(continued - page 2)

Spore taxa*	Macerations				
	152† (%)	2967 (%)	2968 (%)	3118A (%)	902 (%)
Mooreisporites inusitatus					X
Reinschospora magnifica					X
Reticulatisporites reticulatus				0.5	X
R. reticulocingulum				1.0	
Crassispora kosankei			1.0	1.0	X
Granasporites medius					1.0
Densosporites sphaerotriangularis	X			0.5	X
D. triangularis		0.5		X	X
Lycospora granulata	X	61.5	22.0	4.0	8.0
L. micropapillata		4.5			
L. pellucida					0.5
L. pusilla		11.5			
Paleospora fragila					X
Cirratriradites annulatus					0.5
C. saturnii				X	X
Endosporites globiformis	X	X	27.0		
E. plicatus					X
Alatisporites hexalatus	X				
A. punctatus				X	
A. trialatus				0.5	
Laevigatosporites desmoinesensis	X	1.0	1.0	3.5	4.0
L. globosus	X	0.5	23.0	51.0	35.0
L. medius				X	0.5
L. ovalis	X	2.5	4.0	10.0	10.0
L. punctatus				0.5	
L. vulgaris	X			0.5	X
L. sp.				X	
Latosporites minutus		5.0	2.0	6.5	
Puntatosporites minutus		6.5	4.0	3.0	15.5
Spinosporites exiguus		1.5	X	1.0	
Tuberculatosporites robustus				X	
Torispora securis				2.5	X
Vestispora clara					X
V. foveata			X		
V. laevigata			X		X
Florinites mediapudens	X	2.0	9.0	3.0	4.0
F. similis					X
Wilsonites delicatus			X		0.5
W. vesicatus		0.5	2.0		0.5

*Formal systematics of taxa given in Appendix A.
†Relative abundance of spores not determined because of poor preservation.
§X = present but not in count.

TABLE 11. SPORE ANALYSIS OF THE MCALESTER COAL BED IN PONTOTOC COUNTY, OKLAHOMA (MACERATION 2818), AND LOWER BENCH (MACERATION 3153A), AND UPPER BENCH (3153B) OF THE MCALESTER COAL IN PITTSBURG COUNTY, OKLAHOMA

Spore taxa*	Macerations			Spore taxa*	Macerations		
	2818 (%)	3153A (%)	3153B (%)		2818 (%)	3153A (%)	3153B (%)
Punctatisporites minutus	0.5	0.5		*Murospora kosankei*	X		
Calamospora breviradiata	7.0		X†	*Densosporites sphaerotriangularis*	X		
C. hartungiana	0.5	X		*Lycospora granulata*	18.0	30.5	36.0
C. spp.		0.5		*L. micropapillata*		1.0	3.5
				L. orbicula	3.5		
Granulatisporites adnatoides	X			*L. pellucida*	19.0	21.5	18.0
G. granularis		0.5		*L. pusilla*	3.0	2.5	11.0
G. granulatus		0.5					
G. minutus			0.5	*Cirratriradites annulatus*			X
G. verrucosus		0.5		*Endosporites globiformis*	7.0	7.0	1.5
G. sp.		0.5					
				Laevigatosporites desmoinesensis	0.5	8.5	9.5
Cyclogranisporites aureus	X		1.0	*L. globosus*	3.0	13.0	1.0
Lophotriletes ibrahimii	X			*L. medius*	X		
L. microsaetosus	1.0		0.5	*L. ovalis*	11.0	4.0	2.5
L. mosaicus	X			*L. striatus*			X
Pustulatisporites crenatus		0.5					
Pilosisporites aculeolatus	0.5			*Latosporites minutus*	X		
				Punctatosporites minutus	3.5	2.5	1.5
Raistrickia abdita	X		X	*Spinosporites exiguus*	1.0		
R. crocea	X			*Thymospora pseudothiessenii*	X	X	
R. sp.	X			*Torispora securis*	X	X	X
Microreticulatisporites nobilis	X		4.5				
M. sulcatus		X	X	*Vestispora clara*	X		
Camptotriletes sp.	0.5			*V. foveata*	X		
				V. wanlessii	X		X
Triquitrites additus	0.5		0.5				
T. bransonii	2.0	2.0	0.5	*Florinites mediapudens*	14.0	3.5	5.0
T. exiguus	X	X	X	*F. similis*	X		
T. subspinosus	X			*F. sp.*	0.5		
T. sp.			0.5				
Knoxisporites stephanephorus			X				
Crassispora kosankei	X						
Granasporites medius	3.5	1.0	2.0				

*Formal systematics of taxa given in Appendix A.
†X = present but not in count.

eastern Illinois that is correlative with the Dawson Springs (No. 4) and Lewisport coal beds in western Kentucky, but the name was not formally designated for use in Illinois. It is now concluded from palynological data that the O'Nan coal bed, whose name was formally adopted (Kosanke et al., 1960) for a coal in southeastern Illinois, is in the same position as the Mining City, Curlew (Owen, 1856), Mannington, Cates, Dawson Springs, and Lewisport coal beds (Peppers, 1995). Because the name Lewisport coal has priority, it should be used to designate the coal. The Lewisport Coal Bed lies between the Delwood (which correlates with the Bidwell Coal) and Rock Island Coal Beds in Illinois and extends into southeastern Illinois, where it has been identified palynologically, mostly from cores and well cuttings. It also correlates with the Buffaloville Coal Member in southern Indiana (Peppers, 1995).

Appalachian coal region. The Atokan-Desmoinesian boundary is correlated in this study with the base of the Princess No. 5B coal bed in eastern Kentucky, the base of the Upper Mercer coal in Ohio, and the Upper No. 5 Block coal bed in West Virginia.

Rice et al. (1979a) correlated the Atokan-Desmoinesian boundary with the Skyline and Princess No. 5 coal beds, which are correlative, in eastern Kentucky: in the same year (1979b), they correlated it with the top of the Main Ore, which they considered equivalent to the Upper Mercer limestone in Ohio. Arkle et al. (1979) considered the Atokan-Desmoinesian boundary to lie just below the base of the Allegheny Formation in northern West Virginia. The *Neuropteris rarinervis* floral zone of Read and Mamay (1964) in the lower part of the Allegheny Formation in the Appalachian region is similar to the compression flora in the Hartshorne Sandstone in Arkansas and Oklahoma (Hendricks and Read, 1934), which is at the base of the Desmoinesian.

TABLE 12. SPORE ANALYSIS OF THE LOWER (MACERATION 3035B) AND UPPER (MACERATION 3035C) BENCHES OF THE COAL BED THAT HAD BEEN PREVIOUSLY ASSIGNED TO THE BOSTWICK MEMBER IN CARTER COUNTY, OKLAHOMA

Spore taxa*	Macerations 3035B (%)	3035C (%)	Spore taxa*	Macerations 3035B (%)	3035C (%)
Deltoidospora adnata	X†		*Reticulatisporites reticulatus*	X	
D. priddyi	X		*Murospora kosankei*	X	
Punctatisporites minutus	2.5		*Lycospora granulata*	9.0	63.0
			L. micropapillata	X	
Calamospora breviradiata	5.0	2.0	*L. pellucida*	1.5	0.5
C. hartungiana	2.0		*L. pusilla*	1.0	
C. liquida	X				
C. mutabilis	X		*Cirratriradites annuliformis*	X	
C. straminea	1.0		*C. maculatus*	X	
			C. saturnii	X	X
Granulatisporites granularis		X	*Endosporites globiformis*	X	X
G. micrograniter	0.5				
Cyclogranisporites aureus	0.5		*Laevigatosporites desmoinesensis*	1.5	5.5
C. minutus	1.5		*L. globosus*	26.0	7.5
C. obliquus	1.5		*L. medius*		1.0
			L. ovalis	2.0	1.0
Verrucosisporites microtuberosus	0.5		*L. striatus*	0.5	0.5
Lophotriletes commissuralis	0.5		*L. vulgaris*	0.5	0.5
L. ibrahimii	2.0				
L. microsaetosus	3.5	8.5	*Punctatosporites minutus*	1.0	1.0
L. rarispinosus	1.0		*Thymospora pseudothiessenii*	X	
			Torispora securis	9.5	
Anapiculatasporites spinosus		0.5	*Vestispora foveata*	1.5	
Apiculatasporites setulosus	0.5	X	*V. pseudoreticulata*	X	
A. sp.	0.5		*V. wanlessii*	X	
Pilosisporites aculeolatus	7.5	1.5			
P. williamsii	X		*Florinites mediapudens*	5.5	5.5
			F. millotti	1.0	
Raistrickia abdita	X		*F. similis*	X	
R. aculeolata	X		*F. volans*	0.5	
R. crocea	X		*F. spp.*	0.5	0.5
Microreticulatisporites nobilis	0.5	X			
M. sulcatus		X	*Wilsonites delicatus*		X
			W. vesicatus	X	
Triquitrites additus	X		*Pityosporites westphalensis*	X	
T. bransonii	7.5	0.5	*Vesicaspora wilsonii*		0.5
T. exiguus	X	X			
T. sculptilis	X				
T. tribullatus	X				

*Formal systematics of taxa given in Appendix A.
†X = present but not in count.

In the Princess Reserve District of eastern Kentucky, *Radizonates difformis* extends as high as the Skyline coal (Kosanke, 1973). I have also observed the species in the Skyline coal. In the Illinois basin it ranges up to the Pope Creek Coal near the top of the Atokan in Illinois and up to a local unnamed rider coal of the Upper Block Coal Member in Indiana, which is correlated with the Pope Creek Coal. *Renisporites* also ranges as high as the Skyline coal (Kosanke, 1973), and I have observed it from as high as the Pope Creek Coal in Illinois. *Murospora kosankei* first appears in the Princess No. 5 coal in Kentucky and just above the Rock Island Coal near the base of the Desmoinesian in Illinois. Wilson (1976b) reported *Zosterosporites triangularis*

from clastic rocks in the type Atokan. Kosanke (1973) reported the species in the Princess No. 5 coal and its roof rock. I observed it in the Broaz coal bed, which correlates with the Princess No. 4 coal (Rice and Smith, 1980), the Hazard No. 4 coal, and a rider coal of the Hazard No. 4 coal (Table 8). The earliest recorded appearance of *Z. triangularis* in the Appalachian region is in the Coal Creek coal bed in Tennessee (Table 4).

Kosanke (1984) studied the palynology of coal beds in the area of the proposed stratotype of the Pennsylvanian in West Virginia. He indicated that not all of the samples he studied and called Lower No. 5 Block coal bed or Upper No. 5 Block coal

TABLE 13. SPORE ANALYSIS OF THE UNNAMED COAL (MACERATION 2567F) THAT UNDERLIES THE CURLEW LIMESTONE MEMBER IN UNION COUNTY, KENTUCKY, AND IS EQUIVALENT TO THE ROCK ISLAND COAL MEMBER IN ILLINOIS

Spore taxa*	Maceration 2567F (%)
Punctatisporites minutus	9.0
P. sp.	0.5
Calamospora breviradiata	0.5
Granulatisporites adnatoides	0.5
G. granularis	1.0
G. minutus	1.5
G. pallidus	2.0
G. verrucosus	2.0
Cyclogranisporites obliquus	0.5
C. orbicularis	2.0
Pilosisporites aculeolatus	0.5
Raistrickia crocea	X†
Microreticulatisporites sulcatus	0.5
Dictyotriletes bireticulatus	X
Triquitrites bransonii	3.5
T. exiguus	1.0
T. sculptilis	0.5
Granasporites medius	7.0
Lycospora granulata	19.0
L. micropapillata	0.5
L. pellucida	1.5
L. pusilla	1.0
L. rotunda	1.5
Cirratriradites maculatus	X
Endosporites globiformis	0.5
E. zonalis	X
Laevigatosporites desmoinesensis	6.0
L. globosus	13.5
L. ovalis	5.0
L. striatus	0.5
L. vulgaris	X
Latosporites minutus	3.0
Punctatosporites minutus	6.5
Spinosporites exiguus	0.5
Torispora securis	5.0
Vestispora foveata	0.5
Florinites mediapudens	2.5
F. visendus	X
F. volans	0.5

*Formal systematics of taxa given in Appendix A.
†X = present but not in count

bed should be assigned to those coals. A late Atokan age is indicated by the abundance of *Laevigatosporites globosus, Punctatosporites minutus, Torispora securis,* and *Radiizonates,* as well as the presence of *Dictyotriletes bireticulatus* in the Lower No. 5 Block coals and some of the samples of Upper No. 5 Block coal. A significant percentage of *Thymospora pseudothiessenii* in the youngest sample of the Upper No. 5 Block coal indicates a Desmoinesian age. The most diagnostic changes in spore assemblages occur between coals of Kosanke's macerations 554B, 574, 554A, and 573 of the Upper No. 5 Block coal. *Radiizonates* ranges up to the coal of maceration 554B; it doesn't occur in the coal of maceration 574, but reappears once more in the overlying coal of maceration 554A. *Mooreisporites inusitatus* appears for the first time in maceration 574, and *Thymospora pseudothiessenii* first appears in maceration 573. These two species begin their ranges elsewhere at the beginning of the Desmoinesian. The Atokan-Desmoinesian boundary probably correlates with the base of the coal of maceration 574. Kosanke (1988a) later concluded that the coal of maceration 554A may be a little older than the Upper No. 5 Block coal. The samples of the Upper No. 5 Block coal he studied in 1988 are apparently all early Desmoinesian in age. He noted that *Torispora* became very abundant a little below the Upper No. 5 Block coal, and a similar abundance also occurs just below and above the Atokan-Desmoinesian boundary in the Illinois basin. Hower et al. (1992) also noted an abundance of fern spores, including a large percentage of *Torispora securis* in the No. 5 Block coal in northeastern Kentucky, and correlated the coal with the lower part of the Allegheny Formation. The Pope Creek coal in Illinois, which is a little below the Atokan-Desmoinesian boundary, also contains a large percentage of *Torispora* and other fern spores.

Eble and Gillespie (1986a) noted the presence of *Radiizonates* up to the Stockton A coal above the Stockton coal in West Virginia. They correlated the latter with the Rock Island Coal of Illinois, but the Stockton should be a little older than the Rock Island because *Radiizonates* does not extend into the Rock Island Coal. *Zosterosporites triangularis,* which occurrs from the Atokan to the lower Desmoinesian in Illinois, ranges from the Coalburg coal to the Upper No. 5 Block coal in West Virginia (Kosanke, 1988a).

In Ohio, the last occurrence of *Dictyotriletes bireticulatus* is in the Middle Mercer coal, and *Thymospora pseudothiessenii* was first observed in a coal below the Upper Mercer Limestone (Table 14). The Atokan-Desmoinesian boundary therefore lies between those two coals.

Western Europe. The Atokan-Desmoinesian boundary is herein correlated with the Westphalian C-D boundary in western Europe, which is discussed in the following section.

The Atokan-Desmoinesian boundary was considered equivalent to the upper Westphalian C (Peppers, 1984; Phillips and Peppers, 1984; Phillips et al., 1985), the upper part of the Westphalian B (Harland et al., 1990; Klein, 1990), the base of the Westphalian C (Sohn and Jones, 1984), and the late Westphalian A (Ross and Ross, 1985, 1988).

TABLE 14. SPORE ANALYSIS OF THE MIDDLE MERCER COAL BED (MACERATION 2911);
LOWER, MIDDLE, AND UPPER BENCHES OF THE UPPER MERCER COAL BED
(MACERATIONS 2910A-C); THE UNNAMED COAL (MACERATION 2930) BELOW THE
BROOKVILLE COAL BED; THE BROOKVILLE COAL BED (MACERATION 2931); AND THE
CLARION COAL BED (MACERATION 2643) IN OHIO

Spore taxa*	Macerations						
	2911 (%)	2910A (%)	2910B (%)	2910C (%)	2930 (%)	2931 (%)	2643 (%)
Deltoidospora levis		X†					
D. priddyi						X	
D. sphaerotriangula					0.5		
D. sp.							X
Punctatisporites cf. *edgarensis*	0.5				X		
P. flavus							X
P. glaber	0.5		1.0		0.5		
P. minutus	0.5	3.5	34.0	1.5		0.5	1.0
P. obesus	1.0	X		X			X
Calamospora breviradiata							2.0
C. hartungiana							X
C. mutabilis							X
C. straminea					0.5		
C. spp.						0.5	
Granulatisporites adnatoides					X	X	
G. minutus		X		0.5	0.5		
G. pallidus			X		X	X	
G. verrucosus		0.5					
G. spp.		X		0.5	X		
Cyclogranisporites aureus		0.5					
C. cf. *aureus*		0.5			X		
C. leopoldi				0.5	X		
C. microgranus	2.0		X		X		
C. obliquus			11.5	0.5			0.5
C. orbicularis			1.0	0.5			0.5
C. staplinii			0.5				X
Verrucosisporites donarii		X					
V. microtuberosus	0.5		1.0		0.5	X	
V. morulatus	X						
V. sifati	4.0	1.0	X	3.0			X
V. verrucosus					X	X	
V. cf. *verrucosus*	1.5			0.5			X
V. spp.		1.0	X				X
Lophotriletes microsaetosus		1.0					
L. mosaicus		X					
L. pseudaculeatus							X
L. rarispinosus					X		
Anapiculatisporites spinosus		1.0			X		
Pustulatisporites sp.					X		X
Apiculatasporites latigranifer					X	X	
A. spinososaetosus							X
A. sp.		0.5					
Pilosisporites aculeolatus		1.0			X		
P. sp.					X		
Raistrickia abdita					X	0.5	
R. breveminens					X		
R. crinita					X		
R. crocea					X		

TABLE 14. SPORE ANALYSIS OF THE MIDDLE MERCER COAL BED (MACERATION 2911); LOWER, MIDDLE, AND UPPER BENCHES OF THE UPPER MERCER COAL BED (MACERATIONS 2910A-C); THE UNNAMED COAL (MACERATION 2930) BELOW THE BROOKVILLE COAL BED; THE BROOKVILLE COAL BED (MACERATION 2931); AND THE CLARION COAL BED (MACERATION 2643) IN OHIO (continued)

Spore taxa*	Macerations						
	2911 (%)	2910A (%)	2910B (%)	2910C (%)	2930 (%)	2931 (%)	2643 (%)
Microreticulatisporites nobilis	X	X			X		
M. sulcatus	0.5	1.0	X		X		0.5
Dictyotriletes bireticulatus	X						
Camptotriletes confertus					X		
Ahrensisporites guerickei					0.5		
Triquitrites additus					X		0.5
T. bransonii		X	1.5	0.5	1.0		2.0
T. exiguus		X			X		0.5
T. minutus				0.5	X		
T. protensus							0.5
T. pulvinatus					X		
T. sculptilis		1.0			X	X	
T. subspinosus					X		
Zosterosporites triangularis					X		
Mooreisporites inusitatus		0.5					X
Reinschospora triangularis							X
Knoxisporites stephanephorus	X						
Reticulatisporites reticulatus	3.0					0.5	
R. reticulocingulum	X						X
R. sp.				0.5			
Crassispora kosankei					0.5		
Granasporites medius		0.5			2.5	3.0	0.5
Murospora kosankei					X		
Densosporites sphaerot-riangularis		X		0.5			X
D. triangularis		2.0	X	1.0	0.5	X	
D. spp.		1.0			X		
Lycospora granulata		1.0	0.5		22.5	31.5	32.0
L. micropapillata			0.5		1.5		
L. pellucida					18.0	1.5	
L. pusilla		0.5			8.0	4.5	
L. rotunda		0.5			1.0		
Cirratriradites annuliformis						X	
C. maculatus					0.5	X	
C. reticulatus					X		
Endosporites globiformis			0.5				
Alatisporites hoffmeisterii					X		
A. pustulatus							X
A. trialatus	X						
Laevigatosporites desmoinesensis	1.5	1.0			3.5		
L. globosus	11.5	25.5	0.5	54.0	24.0	41.5	31.5
L. medius	3.0			2.5	X		
L. ovalis	41.0	2.0	1.5	4.0	1.5	6.5	2.5
L. punctatus		0.5					7.5
L. vulgaris	0.5	X	X		X	X	

TABLE 14. SPORE ANALYSIS OF THE MIDDLE MERCER COAL BED (MACERATION 2911); LOWER, MIDDLE, AND UPPER BENCHES OF THE UPPER MERCER COAL BED (MACERATIONS 2910A-C); THE UNNAMED COAL (MACERATION 2930) BELOW THE BROOKVILLE COAL BED; THE BROOKVILLE COAL BED (MACERATION 2931); AND THE CLARION COAL BED (MACERATION 2643) IN OHIO (continued)

Spore taxa*	Macerations						
	2911 (%)	2910A (%)	2910B (%)	2910C (%)	2930 (%)	2931 (%)	2643 (%)
Latosporites minutus	1.0	0.5	1.5	5.0			
Punctatosporites minutus	4.0	42.5	36.5	8.5	9.0	8.5	13.5
Spinosporites exiguus	1.0	2.0	2.0	1.5	0.5		
Thymospora obscura							X
T. pseudothiessenii		X					X
Torispora securis	4.0	4.0	2.5	11.0	X	X	X
Vestispora fenestrata					X		
V. laevigata					X	X	
Florinites mediapudens	12.5	2.5	3.5	2.0	0.5	1.0	4.5
F. volans	X	X			2.0	X	
Wilsonites delicatus	3.5						
W. vesicatus	2.5	1.0		1.0	X		
Peppersites ellipticus					X		

*Formal systematics of taxa given in Appendix A.
†X = present but not in count.

Donets basin. The Atokan-Desmoinesian boundary is herein correlated with Limestone M_3 in the lower part of the C_2^7 suite in the Donets basin.

Westphalian C (Bolsovian)-D boundary

Western Europe. The boundary between the Westphalian C and D Stages is herein considered chronostratigraphically equivalent to the Atokan-Desmoinesian boundary.

According to an agreement of the Second Heerlen Congress (1935), the first appearance of the compression fossil *Neuropteris ovata* marks the lower boundary of the Westphalian D. Smith and Butterworth (1967) correlated the Westphalian C-D boundary in Britain with the first appearance of *Thymospora*; however, Grebe (1972), Loboziak (1974), and Van de Laar and Fermont (1990) indicated that *Thymospora* occurs rarely a little before the appearance of *Neuropteris ovata* and the Westphalian C-D boundary in northern France, the Ruhr basin, and The Netherlands, respectively. In their summaries of Carboniferous miospores in western Europe, Clayton et al. (1977) and Owens et al. (1978) therefore indicated that *Thymospora* first appears in uppermost Westphalian C strata. Although the first appearances of *Neuropteris ovata* and *Thymospora pseudothiessenii* are used as references for the beginning of the Westphalian D, some discrepancies in this relation occur. Darrah (1970, p. 105) noted that a problem in identifying the stratigraphic range of *N. ovata* exists because of the "loose usage of the name *ovata* in America but more so because the specific concept has been so amplified during the years to have lost its original meaning."

Other palynological criteria are the same for recognizing the base of the Westphalian D as for the base of the Desmoinesian. *Cadiospora magna* begins its range at the base of the Westphalian D, and *Mooreisporites inusitatus* first appears in the earliest Westphalian D or perhaps in the late Westphalian C in Great Britain (Smith and Butterworth, 1967). *Reticulatisporites polygonalis, Ahrensisporites guerickei* var. *ornatus*, and *Cirratriradites saturni* extend up to about the Westphalian C-D boundary in western Europe (Plate 2). *Savitrisporites nux* disappears a little below, and *Dictyotriletes bireticulatus* disappears a little above the boundary.

Illinois basin. The boundary between the Westphalian C and D is herein correlated with the top of the Seville Limestone Member in northern Illinois, the Curlew Limestone Member in western Kentucky, and the Perth Limestone Member in Indiana.

Bode (1958) found the compression fossil *Neuropteris ovata*, which marks the beginning of the Westphalian D in Europe, in the "horizon" of the Rock Island Coal in northwestern Illinois. Also, it is common in strata associated with the Murphysboro Coal Member, which is several coals younger than the Rock Island Coal. I have also observed several specimens of *Thymospora pseudothiessenii* in the Rock Island Coal. The two species thus appear almost at the same time in the Illinois basin as they do in Britain. *Thymospora* gradually becomes more consistently present in younger coals and is recognized easily in the Murphysboro Coal Bed. The base of the epibole of *Thymospora* is at the Davis, Dekoven, or Seelyville Coal Members at the top of the Tradewater Formation at about the middle of the Westphalian D.

If the precise definition of the base of the Westphalian D is followed, the appearance of *N. ovata* in the "horizon" of the Rock Island Coal makes the Westphalian C-D boundary at about the same age as the overlying Seville Limestone. The term "horizon" may refer to the shale between the coal and overlying limestone or to clastic rocks in approximately the same position as the coal. Because *N. ovata* probably did not appear at exactly the same time in Europe as it did in the Illinois basin, and because palynological evidence supports a latest Westphalian C age for the Rock Island Coal, the Westphalian C-D boundary is correlated with the top of the Seville Limestone.

Cadiospora magna and *Mooriesporites inusitatus* first appear just above the Seville Limestone and at the base of the Westphalian D in Europe. Spore taxa that disappear just below the Seville Limestone and the base of the Westphalian D are *Ahrensisporites guerickei* var. *ornatus*, *Cingulizonates loricatus*, *Cristatisporites indignabundus*, *Cirratriradites saturni*, *Reticulatisporites polygonalis*, *Knoxisporites triradiatus*, *Simozonotriletes intortus*, and *Vestispora magna*. *Savitrisporites nux* disappears a little earlier than those taxa, and *Dictyotriletes bireticulatus* disappears just above the limestone and the Westphalian C-D boundary in Europe.

Western Interior coal province. The base of the Westphalian D is correlated with the base of the Hartshorne Formation in Oklahoma and at the base of the Cliffland Coal in Iowa. If the Cliffland Coal is absent and a limestone is present above the coal, the Westphalian C-D boundary is placed at the top of that. The boundary is at the top of the Riverton Formation in Missouri as currently defined and at the base of the Krebs Formation.

The first appearance of *Neuropteris ovata*, which marks the beginning of the Westphalian D in Europe, was recorded from a composite flora from shale above the Hartshorne coal and the coal at the base of the overlying McAlester Formation in Oklahoma (Hendricks and Read, 1934). *Thymospora*, which is also an index fossil for the base of the Westphalian D, is apparently less abundant in the Western Interior coal province (Wilson, 1976b) than it is in the Illinois basin and Europe. *Thymospora* was reported in the Drywood coal in Oklahoma by Bordeau (1964) and Wilson (1976b); however, Urban (1965) did not observe the species in his samples of the Drywood coal, nor did I in my sample (Table 10) from Missouri. The Drywood coal is above the McAlester Formation. The Rowe coal, which is between the McAlester and Drywood coals, was investigated by Davis (1961). He compared its spore assemblages with those in the Pope Creek, Tarter, and Manley Coals in Illinois because the Rowe coal had been correlated earlier with the Tarter Coal (Kosanke et al., 1960). Palynological evidence indicates that the Drywood and Rowe coals are correlative with early Westphalian D and are younger than the Pope Creek and Tarter Coals.

Morgan (1955a, 1955b) and Dempsey (1964, 1967) did not observe *Thymospora pseudothiessenii* in the McAlester coal in Oklahoma. I observed several specimens of *T. pseudothiessenii* in samples of the McAlester coal in Pontotoc County (maceration 2818, Table 11) and Pittsburg County (maceration 3153

A-B). No spores restricted to strata older than Westphalian D were observed.

In Iowa, the regular occurrence of *Thymospora pseudothiessenii* begins in a coal at about the middle of the Floris Formation, which is probably equivalent to the Davis Coal in Illinois (Ravn, 1986). Ravn (1986) correlated the Westphalian C-D boundary with the base of the Whitebreast Coal Member, which is equivalent to the Colchester Coal Member and a little above the Davis Coal at the base of the Carbondale Formation in Illinois. The Colchester Coal and the middle of the Westphalian D in Europe are near the base of the epibole of *Thymospora*. Ravn (1986) observed some rare specimens comparable to *T. pseudothiessenii* from coals as old as the Blackoak and Laddsdale Coal Members, but he expressed doubt concerning their identification. The specimens of *T.* cf. *pseudothiessenii* from the Blackoak Coal illustrated by Ravn (1979), however, appear nearly identical to specimens of *T. pseudothiessenii* illustrated by him (1986) from younger coals. As mentioned previously, I observed rare specimens of *T. pseudothiessenii* in the Rock Island Coal in Illinois, which is slightly younger than the Blackoak Coal. The scarcity of *T. pseudothiessenii* at the beginning of its range makes it difficult to use for making routine biostratigraphic correlations, but it still can be an aid in indicating the beginning of the Westphalian D.

Ravn (1986) observed specimens of *Cadiospora magna*, which first appears at the base of the Westphalian D in Europe and Illinois (Plate 2), in a coal thought to belong to the Laddsdale Coal complex. Ravn (1986) interpreted the Laddsville Coal as belonging to the lower Westphalian C, but I consider it to be early Westphalian D in age. *Cadiospora magna* regularly occurs in and above the Wheeler Coal Member, which I include in the middle Westphalian D. *Mooreisporites inusitatus* first appears in an unnamed coal that Ravn (1986) thought to be equivalent to the Cliffland Coal, and appears at the base of the Westphalian D in Europe and Illinois. Coquel (1976) reported the species in coals as young as Westphalian A; however, it may be a poorly preserved specimen of the older *Mooreisporites trigallerus* or *M. fustus*. *Savitrisporites nux* and *Cristatisporites indignabundus* range up to the Blackoak Coal and to the basal part of Westphalian D in Europe. *Knoxisporites triradiatus* and *Simozonotriletes intortus* are also present in the Blackoak Coal and range up to the Westphalian C-D boundary. *Reticulatisporites polygonalis* and *Alatisporites pustulatus* extend up to the Cliffland Coal, whereas the former occurs up to at least the top of the Westphalian C (Coquel, 1976) and the latter up to lowermost Westphalian D.

Appalachian coal region. The Westphalian C-D boundary correlates palynologically with the base of the Princess No. 5B coal bed in eastern Kentucky, the base of the Upper Mercer coal bed in Ohio, and the Upper No. 5 Block coal bed in West Virginia.

Gillespie and Pfefferkorn (1979) correlated, by use of plant compressions, the Westphalian C-D boundary with the upper part of the Kanawha Formation in the area of the USGS proposed Pennsylvanian stratotype. The Westphalian C-D was correlated with the base of the Allegheny Formation in the Appalachian region by Phillips and Peppers (1984) and with the top of the Upper Mercer coal (Phillips et al., 1985).

Bode (1975, p. 148) stated that, in eastern Kentucky, "the base of the zone of *Neuropteris ovata* must be between the Princess No. 5 and Skyline seams," which are now considered correlative (Rice and Smith, 1980). Because *Thymospora pseudothiessenii* was not reported in the overlying Princess No. 5A coal (Kosanke, 1973), but in the seat rock of the Princess No. 5B coal, *N. ovata* may have first appeared slightly earlier than *T. pseudothiessenii* in eastern Kentucky. Kosanke (1973) reported 1% to 6% *T. pseudothiessenii* in the Princess No. 6 coal bed. I examined a sample of the Princess No. 6 coal from Boyd County, Kentucky, that contains 26% *T. pseudothiessenii*. That species first appears a little above the middle of the Tradewater Formation in the Illinois basin (Peppers, 1984), but it becomes abundant much higher, near the top of the formation. Therefore, the interval from the Princess No. 5B coal to the Princess No. 6 coal in Kentucky is probably equivalent to most of the upper half of the Tradewater Formation in Illinois.

In Ohio, *Thymospora pseudothiessenii* first appears in the Upper Mercer coal (Table 14). The overlying Brookville coal at the base of the Allegheny Formation and even younger Clarion coal also contain rare specimens of the species and an abundance of *Laevigatosporites globosus* and other fern spores, which first become abundant in the upper part of the Westphalian C in Illinois.

According to White (1900), the flora associated with the Stockton coal in West Virginia belongs to an Allegheny type of flora, probably similar to that in the "Clarion Group" of western Pennsylvania. Jongmans and Gothan (1934) also concluded that the zone of *N. ovata* begins in the lower part of the Allegheny Formation, and this agrees with palynological evidence. Read and Mamay's (1964) floral zone 8 in West Virginia, which includes all of the Kanawha Formation except for the lower part, marks the first appearance of *Neuropteris* of the *N. ovata* Hoffman type. Darrah (1970) reported that *N. ovata* begins its range in the upper part of the Kanawha Formation, and Gillespie and Pfefferkorn (1979) stated that, in the area of the proposed Pennsylvanian stratotype, the Westphalian D begins just above the Stockton coal at the first occurrence of *N. ovata*.

As mentioned in the previous section, *Thymospora pseudothiessenii* was recorded by Kosanke (1984) as first appearing at 12% of the spore assemblage in an upper bench (maceration 573-C) of the Upper No. 5 Block coal in West Virginia, a coal that is a little younger than the Stockton coal. In Illinois and Europe, the species does not become common until about the middle of the Westphalian D. Kosanke (1984) tentatively correlated the No. 6 Block coal with the Lower Kittanning coal of Ohio and Pennsylvania, because it contains *Schopfites dimorphus* and abundant *T. pseudothiessenii* (Gray, 1967). Eble and Gillespie (1986b) observed one questionable specimen of *T. pseudothiessenii* in the Coalburg coal, which was sampled at 11

localities in West Virginia. I do not yet recognize this singular occurrence as the beginning of the range of *Thymospora* because it has not been reported in the overlying Stockton, Stockton A, or Lower No. 5 Block coal (Kosanke, 1988a). Also, the presence of *Dictyotriletes bireticulatus, Radiizonates* cf. *difformis,* and *Densosporites annulatus* in the Coalburg coal indicates a late Westphalian C age. Conversely, the occurrence of *Neuropteris ovata* just above the Stockton coal (Gillespie and Pfefferkorn, 1979) places its appearance before the well-documented first appearance of *Thymospora*, the opposite of their relation in Europe.

Mooreisporites inusitatus appears for the first time in the lower bench of the Upper No. 5 Block coal in West Virginia (Kosanke, 1988a) and at the base of the Westphalian D in Britain (Smith and Butterworth, 1967). The Coalburg coal marks the last appearance of *Cristatisporites indignabundus, Savitrisporites nux, Ahrensisporites guerickei,* and *Knoxisporites triradiatus* (Kosanke, 1988a; Eble and Gillespie, 1986b). The Lower No. 5 Block coal bed is the youngest coal containing *Cirratriradites saturni* and *Radiizonates difformis*. All of these taxa disappear just below the Westphalian C-D boundary. Thus, as in the Princess reserve district, the lower half of the Westphalian D strata in West Virginia between the Lower and Upper No. 5 Block coals is greatly reduced in thickness compared with that in Europe and the Illinois basin.

Donets Basin. Limestone M_3 in the lower part of the C_2^7 suite in the Donets basin marks the Westphalian C-D boundary.

Teteryuk (1976) correlated the Westphalian C-D boundary with Limestone M_1 because *Thymospora* first appears in coal near there. Bouroz et al. (1978) indicated that the boundary is slightly higher at Limestone M_3, which is at about the first appearance of *Neuropteris ovata*. Teteryuk (1982) later agreed with this position. Owens et al. (1978), in comparing Carboniferous palynological zonation in Europe and the Donets basin, stated that *Thymospora* appears in both regions in the upper Westphalian C before *N. ovata* appears. Owens (1984) later stated, however, that *Thymospora* first occurs at the base of the Westphalian D in western Europe and the Donets basin.

Inosova et al. (1975, 1976) also showed that *Thymospora* and *Cadiospora* first appear near Limestone M_1. *Torispora* and *Triquitrites sculptilis* increase in abundance at about the same time as they do in Illinois and in Iowa. *Thymospora* increases significantly at the Limestone N_1 at the base of the C_3^4 suite, which correlates with about middle Westphalian D and the Colchester Coal in Illinois. *Vestispora magna* disappears near the top of suite C_2^6 and the end of the Westphalian C in western Europe.

Middle-Upper Pennsylvanian boundary (USGS) and Moscovian-Kasimovian Series boundary

Appalachian coal region. The Middle-Upper Pennsylvanian boundary proposed by the USGS is herein considered to be approximately equivalent to the Moscovian-Kasimovian boundary. These boundaries are just below the Middle-Upper Pennsylvanian boundary as used in the Midcontinent.

Most stratigraphers in the Appalachian coal region (e.g., Rice et al., 1979b; Gillespie and Pfefferkorn, 1979; Henry et al., 1979; Rice and Smith, 1980; McDowell et al., 1981; Englund et al., 1986; Kosanke, 1988c) followed the proposal of Bradley (1956) and assigned the Allegheny-Conemaugh boundary as the Middle-Upper Pennsylvanian boundary. In the area of the Pennsylvanian stratotype proposed by the USGS, the Middle-Upper Pennsylvanian boundary correlates with the top of the Charleston Sandstone. In Pennsylvania and Ohio, it correlates with the top of the Upper Freeport coal (Collins, 1979). Spore assemblages in the Upper Freeport coal at the top of the Allegheny Formation and in the Mahoning coal in the lower part of the Conemaugh Formation are not greatly different from each other, which is in contrast to the changes that occurred between the Mahoning and overlying coals.

Donets basin. Examination of the chart of stratigraphic ranges and relative abundance of Carboniferous and Permian spores in the Donets basin (Inosova et al., 1975) reveals clearly that the boundary between the Moscovian and Kasimovian Stages at Limestone N_2 in the middle of the C_3^1 suite is just below the major floral change at the Westphalian-Stephanian boundary (Limestone N_4).

Phillips and Peppers (1984), Phillips et al. (1985), as well as others have correlated the Moscovian-Kasimovian and Westphalian-Stephanian boundaries with each other. Wagner and Higgins (1979) and Wagner (1984) used plant compressions to correlate the Moscovian-Kasimovian boundary with the lower Stephanian or middle Cantabrian. No marked change in spore assemblages is indicated at the Moscovian-Kasimovian boundary in the Donets basin. The most abundant spores in coals above and below the boundary are *Lycospora pusilla, Laevigatosporites vulgaris, Calamospora, Densosporites, Cyclogranisporites, Endosporites, Punctatosporites minutus,* and *P. oculus* (Inosova et al., 1975).

Illinois basin. The Middle-Upper Pennsylvanian boundary in the Appalachian region and Moscovian-Kasimovian boundary are correlated with the top of the Danville (No. 7) Coal Member near the base of the Shelburn Formation in Illinois (formerly Modesto Formation) and Indiana, and at the top of the Baker coal bed (No. 13) at the base of the Shelburn Formation (formerly Sturgis Formation) in western Kentucky.

In the Illinois basin, Wanless (1956) and Shaver et al. (1970) correlated the top of the Allegheny Series with the top of the Danville Coal. The Allegheny-Conemaugh (Middle-Upper Pennsylvanian) boundary was correlated with the top of the Carbondale Formation (Kosanke et al., 1960; Phillips et al., 1985). Rice et al. (1979b) correlated the boundary with the top of the Chapel (No. 8) Coal Member in the overlying Shelburn Formation.

Lycospora granulata, Granasporites medius, Punctatosporites minutus, and *Thymospora pseudothiessenii* are the most abundant spore species both above and below the top of the Carbondale Formation (Fig. 2). A large number of spore species dis-

appeared from coal beds between the Danville Coal and the next younger coal, the Rock Branch Coal Member (Fig. 4), which is much thinner and more local in extent than the Danville Coal.

Western Interior coal province. The Middle-Upper Pennsylvanian boundary of the Appalachian region is correlated with the top of the Altamont Limestone in the Marmaton Group of

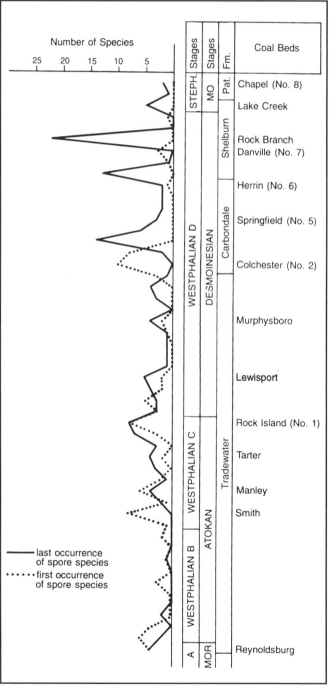

Figure 4. Number of species of spores that first appeared or that disappeared from Middle and Late Pennsylvanian coal beds in the Illinois basin (MO = Morrowan).

Kansas, Iowa, and Missouri, but this is a very tentative correlation because of inadequate palynological data.

Middle-Upper Pennsylvanian Series boundary (Desmoinesian-Missourian Stage boundary)

In the Euramerican equatorial coal belt, a major change in faunas and coal-swamp vegetation occurred simultaneously or nearly so at the Desmoinesian-Missourian boundary (Middle-Upper Pennsylvanian Series boundary) in the Western Interior coal province, Utah, Texas, and the Illinois basin (Fig. 2), just above the Middle-Upper Pennsylvanian boundary in the Appalachian coal region (Kosanke, 1947, 1950; Phillips and Peppers, 1984; Phillips et al., 1985; Peppers, 1984, 1986; DiMichele et al., 1985; Kosanke and Cecil, 1989, 1992; Cecil, 1990; Boardman et al., 1990) and in the Donets basin (Teteryuk, 1974). This change also occurred at the Westphalian-Stephanian Series boundary in western Europe (Fig. 5).

Marked changes in sedimentation and petrographic and mineral composition of coals occurred at the Middle-Upper Pennsylvanian boundary. Schutter and Heckel (1985) cited the upward increase in number of evaporites, caliche horizons, and incompletely leached mixed-layer clays in the Western Interior coal province as evidence for a climate that was becoming pro-

gressively drier from the Desmoinesian to the Permian. An increase in the thickness of regressive limestones deposited at higher sea-level stands during the Missourian indicates less influx of detritus, suggesting a drier climate. In the Appalachian region, a seasonally dry climate during the Late Pennsylvanian is indicated by the occurrence of redbeds, gilgai paleosol structures, calcium carbonate concretions, calcareous shale and sandstone, and nonmarine shallow water limestones, but the climate was not yet dry enough to produce evaporites (Donaldson et al., 1985; Cecil et al., 1985). Grady (1983) and Harvey and Dillon (1985) characterized coals in West Virginia and Illinois, respectively, that were formed under a seasonally dry climate during the late part of the Pennsylvanian. The coals were described as being high in inertinite and low in vitrinite because of exposure of the peat to oxidation. According to Cecil (1990), the eastern United States during the Middle Pennsylvanian had a wet-dry tropical to humid subtropical climate that produced domed peat swamps. During the Late Pennsylvanian, however, the climate became humid subtropical to semiarid and led to the formation of planar peat swamps.

The upper part of the Allegheny Formation and the lower part of the Conemaugh Formation in the Appalachian region as well as the upper part of the Desmoinesian in the Midcontinent are included in floral zone 10 (*Neuropteris flexuosa*) of Read and Mamay (1964). They referred to it as the zone of appearance of abundant species of *Pecopteris*. A flora from near Onaga (Kansas) is the only one in the Midcontinent that represents the overlying floral zone 11 (*Lescuropteris* spp.) in the Missourian (Read and Mamay, 1964).

Wagner (1984, p. 122), who subdivided the Carboniferous of Euramerica into 16 compression floral zones, stated that the base of the *Lobatopteris vestita* zone "is determined by one of the more noticeable changes in the Pennsylvanian." This zone is within the upper Westphalian D and the lower part of the Cantabrian Stage of the Stephanian. Wagner stated that the diversification of pecopterid fern compressions begins with the late Westphalian D, although it is not before the late Cantabrian that the floral assemblages truly become dominated by the pecopterids. Wagner (1984) recognized the well-known Mazon Creek flora overlying the Colchester Coal in Illinois as being in the *L. vestita* zone and at the Westphalian D-Cantabrian transition.

Pfefferkorn and Thomson (1982) showed that compression fossils of ferns greatly increased in relative abundance from the beginning through the middle of the Westphalian D. Others, including Dix (1934), Corsin et al. (1968), Laveine (1976), Wagner (1984), and Cleal (1984), recognized major changes in compression floras that included the addition of new fern taxa in the middle of the Westphalian D. In fact, Zodrow (1986) distinguished two periods of additions of new forms in the middle of the Westphalian D.

The percentage of fern spores in coals increased gradually from the middle to the end of the Atokan (Peppers, 1979), at which time they became very abundant (Figs. 2 and 5). Fern spores became as common as lycopod spores in many of the

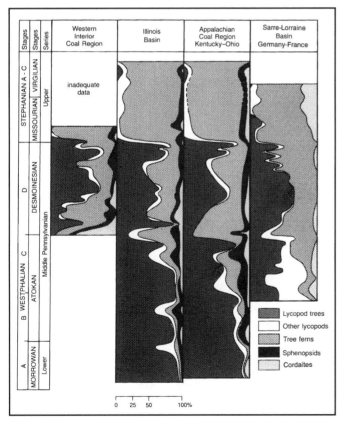

Figure 5. Generalized distribution (in percent) of major plant groups as interpreted from palynological studies of Pennsylvanian coals in the Western Interior coal province, Illinois basin, and Appalachian coal region, and Middle and Upper Carboniferous coals in Europe.

Desmoinesian coals, but decreased in abundance in several coal beds just before the end of the Desmoinesian. Fern spores did not overwhelmingly dominate spore assemblages until the beginning of the Missourian. Note that palynology indicates that the first major increase in ferns in coal swamps was during the late Atokan, whereas studies of plant compressions in clastic rocks indicate that the increase was slightly later, during the early to middle Desmoinesian. The beginning of the increase in abundance of tree-fern spores took place at the same time as the number of species that appeared or disappeared increased (Fig. 4). The decline in the number of first appearances of new spore taxa began in the middle of the Desmoinesian.

Lycospora, which were produced by arborescent lycopods, generally dominate spore assemblages in most Lower and Middle Pennsylvanian coals, but they disappear or greatly diminish in importance in Upper Pennsylvanian coals. Occasional specimens of *Lycospora* are observed in Upper Pennsylvanian coals in North America, but some of these may have been redeposited. *Lycospora* did not disappear, but it is much less abundant in Stephanian limnic basins of central France and in the Donets basin than in the Westphalian. *Granasporites* from *Diaphorodendron* (DiMichele, 1985), another arborescent lycopod, also disappeared at the boundary, except for its occurrence in several early Stephanian coals in France (Alpern, 1959). *Thymospora pseudothiessenii*, a tree-fern spore that was abundant in the Middle Pennsylvanian, declines in importance a few cyclothems before the end of the Desmoinesian and disappeared abruptly at the end of the Desmoinesian. The species is absent through the Missourian in North America, but it is rare in some Virgilian coals in which *T. thiessenii* is common. Other tree-fern spores (e.g., *Laevigatosporites globosus*) also disappeared, and *Punctatosporites minutus* declined sharply in abundance.

Upper Pennsylvanian coals are characterized by an increase in abundance of small tree-fern spores such as *Punctatisporites minutus, Latosporites minutus, Spinosporites exiguus,* and several species of *Cyclogranisporites*. The following species also became more abundant in the Upper Pennsylvanian: *Crassispora* borne by *Sigillaria; Endosporites,* whose parent plant is *Chaloneria* (formerly *Polysporia*); *Calamospora,* which has its affinities with sphenopsids; *Florinites,* produced by cordaites; and other gymnospermic pollen. Their abundance, however, varies greatly from coal bed to coal bed (Fig. 2). The low-growth stature of *Chaloneria* indicates that it grew in shallow-water marshes that were frequently exposed to drying (DiMichele et al., 1979). Other lycopods (e.g., *Sigillaria, Selaginella,* and the parent plant of *Sporangiostrobus,* possibly *Bodeodendron*; Wagner and Spinner, 1976; Wagner, 1989) that survived into the Upper Pennsylvanian and Stephanian were adapted to drier, more clastic environments than were the large tree lycopods (e.g., *Lepidophloios, Diaphorodendron,* and *Paralycopodites*).

Genera and species among different faunal groups also underwent extinctions, appearances, rapid changes in evolution, and changes in diversity at the Desmoinesian-Missourian boundary (Boardman et al., 1990). For example, the ammonoid

Wellerites became extinct a little below the boundary, and *Gonioglyphioceras* has not been found above the boundary (Ramsbottom and Saunders, 1984). *Pennoceras* appeared just above the boundary (Boardman and Mapes, 1984; Boardman et al., 1989c). Among fusulinds, *Fusulina (Beedeina)* became extinct, *Eowaeringella* appeared near the boundary, and *Triticites* appeared a little above the base of the Missourian (Stewart, 1968; Boardman et al., 1990). Conodonts *Neognathodus* and *Gondolella magna* disappeared, and important changes in the evolution of *Idiognathodus* took place at the Desmoinesian-Missourian boundary in the Midcontinent (Swade, 1985; Barrick and Boardman, 1989; Grayson et al., 1989). The occurrence of the brachiopod *Mesolobus* only below the Missourian has been known for a long time (Dunbar and Condra, 1932). Christopher et al. (1990) demonstrated the occurrence of a distinct change in ostracode faunas at the Desmoinesian-Missourian boundary in the Appalachian basin. *Tetractinomorph chaetetids* were major reef builders during the Carboniferous, but declined suddenly in abundance at the end of the Westphalian (West and Archer, 1992).

Article 67 of the North American Stratigraphic Code states that "Boundaries of chronostratigraphic units should be defined in a designated stratotype on the basis of observable paleontological or physical features of the rocks. . . . One means of minimizing or eliminating problems of duplication or gaps in chronostratigraphic succession is to define formally as a point-boundary stratotype only the base of the unit." Article 68, on correlation, states that "Boundaries of chronostratigraphic units can be extended only within the limits of resolution of available means of chronocorrelation, which currently include paleontology, . . . and such indirect and inferential physical criteria as climatic changes, degree of weathering, and relations to unconformities."

According to these guidelines, paleontologists would define the base of the Upper Pennsylvanian by the first appearance of one or more fossil taxa. In the Midcontinent, several diagnostic conodonts and ammonoids appear in the Exline Limestone or equivalent limestones a little above the base of the Pleasanton Group, which has served as the traditional Middle-Upper Pennsylvanian boundary. The most pronounced paleobotanical change in the Pennsylvanian indicated by coal palynology would occur in the lower Upper Pennsylvanian at the first coal above the recognized Middle-Upper Pennsylvanian rather than at the boundary. Recognition of the paleobotanical changes is determined by the disappearance of major lycopod trees and other plants and the corresponding domination of fern taxa, not on the first appearance of new species in the Upper Pennsylvanian. For practical purposes, I am still considering the base of the Pleasanton Group as the Middle-Upper Pennsylvanian boundary. A significant climate change probably did occur at or very close to the boundary. This criterion and palynological and paleontological changes comply with the Stratigraphic Code in defining chronostratigraphic boundaries and making correlations.

Western Interior coal province. Keyes (1893) was the first to refer to the "Des Moines Beds" or "Lower Coal Measures"

for strata along the Des Moines River in Iowa. He stated that the "Lower Coal Measures," which are predominantly argillaceous and contain numerous well-developed coal beds, are well distinguished from the "Upper Coal Measures," which are predominantly calcareous. The top of the Desmoinesian was not specifically defined. The Missourian was first designated a stage by Keyes (1893) for rocks in parts of northwestern Missouri; in 1896, he redesignated it a series. The base of the series was lowered by Moore (1932) from what is now considered Virgilian to the base of the Pleasanton Formation. Moore (1944) recognized the Desmoinesian as one of the series in the Midcontinent time-rock divisions.

At the Lawrence Conference in 1947, the state geological surveys in the Midcontinent region agreed to establish the base of the Pleasanton Formation of Missouri as the base of the Missourian Series (Moore, 1948; Greene and Searight, 1949). The base of the formation is marked by a faunal break and what was thought to be a major unconformity. Cline (1941) and Cline and Greene (1950), however, regarded an unconformity higher in the Pleasanton as being the most important stratigraphic break. Kosanke (cited *in* Howe, 1982) studied the palynology of several samples of coal in the Pleasanton Formation and found that the Locust Creek coal (formerly upper Pleasanton coal; Howe, 1982), is above the range of *Lycospora* and *Thymospora pseudothiessenii.* Thus, a major floral break was thought to coincide with the upper unconformity. Physical and floral evidence indicated to Howe (1982) that the base of the Missourian could be placed at the upper unconformity in the Pleasanton Formation, but that the faunal break above the unconformity at the base of the Pleasanton warranted placing it there. If the Hepler Sandstone at the base of the Pleasanton is missing, the Exline Limestone Member would serve as the base of the Missourian (Howe, 1982).

Palynology demonstrates that a major floral break occurs in Missouri between the Desmoinesian Laredo coal and the Missourian Grain Valley coal (Plate 4, Tables 15–17). The sample of the Laredo coal I studied is from its type section (Howe, 1953) at the top of the Memorial Formation and below the Lost Branch Formation. The Grain Valley coal is equivalent to the Lower Pleasanton coal and overlies the Hepler Sandstone at the base of the Pleasanton Group. *Thymospora pseudothiessenii* greatly dominates the spore assemblage in the Laredo coal (Table 15), and *Lycospora* is well represented. They were not observed in the Grain Valley and younger coals. *Cyclogranisporites* is not common in the Laredo coal, but it is the most abundant genus in the Grain Valley coal. *Punctatisporites minutus* is rare in the Laredo coal, but makes up about three-fourths of the populations in the Locust Creek and Ovid coals overlying the Grain Valley coal (Table 17). Sphenopsid spores (*Calamospora* and large species of *Laevigatosporites*) are a significant part of the spore assemblages in the Grain Valley coal.

Representative samples of coal beds in Kansas that span the Desmoinesian-Missourian boundary (Plate 4, Tables 15–17) also demonstrate that a major floral change occurs

between the Dawson coal and the "Hepler" coal. Lycopsid spores, i.e., *Lycospora granulata, Crassispora kosankei,* and *Granasporites medius (Cappasporites distortus* of earlier reports), constitute more than three-fourths of the spore assemblages in the Desmoinesian unnamed coal (maceration 2805) above the Worland (Altamont) Limestone Member in Kansas (Table 16). An unnamed coal 15 to 20 ft (4.6 to 6.1 m) below the Sni Mills Limestone Member, which is below the Dawson coal, has a spore assemblage that is dominated by *Lycospora granulata* and *Punctatosporites minutus. Lycospora* is well represented in macerations 2809A and 2796 of the Dawson coal, and *Punctatosporites* is about twice as abundant in maceration 2796 as in maceration 2809A. *Lycospora, Granasporites,* and *Crassispora* are not present in the "Hepler" coal above the Desmoinesian-Missourian boundary, but *Endosporites* becomes a significant part of the assemblage. Among the tree-fern spores, *Thymospora* disappears, and the percentage of *Punctatosporites minutus* is greatly reduced in macerations 2809B and 2808 in the Hepler coal. *Cyclogranisporites* becomes dominant. Palynology of clastic rocks between the Dawson and "Hepler" coals in Kansas has shown that the floral change actually occurs a few feet below the base of the Pleasanton Formation (Peppers, 1994). *Lycospora* almost disappears while fern spores and seed-fern pollen (*Wilsonites*) increase greatly in abundance.

In Oklahoma, the Desmoinesian-Missourian boundary is at the base of the Seminole Formation, which correlates with the base of the Pleasanton Group in Kansas and Missouri (Fig. 6). The Tulsa Sandstone Member at the base of the Seminole Formation is equivalent to the Hepler Sandstone in Kansas and Missouri and is unconformable at its base. The Tulsa coal (formerly middle Seminole coal) overlies the Tulsa Sandstone and appears to be stratigraphically and palynologically equivalent to the "Hepler" coal. The "Checkerboard" coal (formerly upper Seminole coal) is between an upper Seminole sandstone and the Checkerboard Limestone and is probably equivalent to the Grain Valley coal. Ravn (1981) reported that the Checkerboard Limestone is the oldest widespread marine stratum of Missourian age in Oklahoma. The Seminole Formation is underlain in east-central Oklahoma by the Desmoinesian Holdenville Formation, which includes the Lost Branch equivalents at its top.

Wilson and Venkatachala (1968) recorded an abundance of *Thymospora pseudothiessenii* and *Lycospora* in the Dawson coal in the Desmoinesian Holdenville Formation. Wilson (1979a, 1979b, 1984) and Wilson and Bennison (1981) discussed the palynology and plant fossils in the Seminole Formation. Wilson (1979a, 1979b) found 1% *Lycospora* in the Tulsa coal and indicated that the Desmoinesian-Missourian boundary should be placed above it. This followed Kosanke's (1950) interpretation that the boundary is at the last occurrence of *Lycospora.* Wilson (1984) also thought that the Tulsa coal is Desmoinesian in age because *Trivolites, Latipulvinites,* and *Centonites,* which occur in some Missourian coals in Illinois (Peppers, 1964), are absent in the Tulsa coal but present in the

TABLE 15. SPORE TAXA IN COAL SAMPLES FROM BELOW AND ABOVE DESMOINESIAN-MISSOURIAN BOUNDARY ILLUSTRATED IN PLATE 4. OKLAHOMA: DAWSON (MACERATION 2715) AND TULSA (MACERATION 2797B) COALS; KANSAS: DAWSON (MACERATION 2809A) AND HEPLER (MACERATION 2809B) COALS; MISSOURI: LAREDO (MACERATION 2780) AND GRAIN VALLEY (MACERATION 2830) COALS; ILLINOIS: POND CREEK (MACERATION 693D) AND LAKE CREEK (MACERATION 693C) COAL MEMBERS; INDIANA:PIRTLE COAL MEMBER (MACERATION 2764) AND UNNAMED COAL (MACERATION 831); OHIO: MAHONING (MACERATION 2789) AND MASON (MACERATION 3101C) COAL BEDS

Spore taxa[†]	Oklahoma		Kansas		Missouri		Illinois		Indiana		Ohio	
	Des. 2715 (%)	Mo. 2797B (%)	Des. 2809A (%)	Mo. 2809B (%)	Des. 2780 (%)	Mo. 2830 (%)	Des. 693D (%)	Mo. 693C (%)	Des. 2764 (%)	Mo.* 831 (%)	Des. 2789 (%)	Mo. 3101C (%)
Deltoidospora grandis										X[§]		
D. levis												X
D. priddyi		0.5			0.5	X						0.5
D. sphaerotriangula			0.5									
D. spp.			0.5									
Punctatisporites flavus					X					X		
P. minutus		2.5	0.5	4.5	0.5	7.5	2.5	1.5	2.0	X	3.0	
P. obesus					X							
P. spp.	0.5											
Calamospora breviradiata		4.0	3.0	2.5			1.5	6.5	1.0	X	1.5	
C. hartungiana	X	1.0				2.0	X			X		7.0
C. pedata							X					
C. straminea						0.5				X		
C. spp.		3.0	0.5			12.5						
Granulatisporites adnatoides						0.5					X	
G. granularis	X		0.5									
G. minutus					0.5	1.5	X			X	X	
G. piroformis							X					0.5
G. spp.						0.5						
Cyclogranisporites aureus		0.5									1.0	
C. microgranus					1.0			X				
C. minutus						1.5						
C. obliquus	X	49.5		59.0		8.0		6.0		X	8.5	
C. orbicularis		4.0	3.5	5.5	1.5	18.0	2.0	26.5		X	2.0	
C. spp.			0.5	1.0		7.5						
Verrucosisporites spp.			X						X			0.5
Lophotriletes sp.									0.5			
L. sp.									X			
Apiculatasporites latigranifer		3.5	1.5	1.0			X	4.5		X		
A. setulosus		0.5	X	0.5				X				
A. spinulistratus			X			X						
A. sp.						0.5						
Raistrickia abdita			X		X		X					
R. aculeata	X						X				X	
R. crocea			X							X		
R. subcrinita			X		X							
R. superba									X			
R. spp.			0.5							X		
Convolutispora fromensis									X			
Microreticulatisporites nobilis		X		X	X					X		0.5
M. sulcatus						0.5				X		X
Triquitrites additus		2.0	0.5	1.0		2.0				X		3.0
T. bransonii	0.5	1.0	1.0	4.0		7.0	1.0	1.5		X	X	6.5
T. crassus					X		X					
T. exiguus											X	
T. spinosus	X				0.5	X	X			X	X	
T. subspinosus				X								

TABLE 15. SPORE TAXA IN COAL SAMPLES FROM BELOW AND ABOVE DESMOINESIAN-MISSOURIAN BOUNDARY ILLUSTRATED IN PLATE 4. OKLAHOMA: DAWSON (MACERATION 2715) AND TULSA (MACERATION 2797B) COALS; KANSAS: DAWSON (MACERATION 2809A) AND HEPLER (MACERATION 2809B) COALS; MISSOURI: LAREDO (MACERATION 2780) AND GRAIN VALLEY (MACERATION 2830) COALS; ILLINOIS: POND CREEK (MACERATION 693D) AND LAKE CREEK (MACERATION 693C) COAL MEMBERS; INDIANA:PIRTLE COAL MEMBER (MACERATION 2764) AND UNNAMED COAL (MACERATION 831); OHIO: MAHONING (MACERATION 2789) AND MASON (MACERATION 3101C) COAL BEDS (continued)

Spore taxa†	Oklahoma		Kansas		Missouri		Illinois		Indiana		Ohio	
	Des. 2715 (%)	Mo. 2797B (%)	Des. 2809A (%)	Mo. 2809B (%)	Des. 2780 (%)	Mo. 2830 (%)	Des. 693D (%)	Mo. 693C (%)	Des. 2764 (%)	Mo.* 831 (%)	Des. 2789 (%)	Mo. 3101C (%)
Mooreisporites inusitatus			X		X							
Reticulatisporites reticulatus					X							
Crassispora kosankei	X		2.0				2.5		8.0		3.5	0.5
Granasporites medius	4.5		17.5				5.0		1.5		1.0	
Lycospora granulata	69.0		23.0		13.0	0.5	39.5	X	52.0		58.5	
L. micropapillata			0.5						13.5			
L. pusilla						X						
L. rotunda			1.0				X					
Cirratriradites annulatus									X			
C. annuliformis	X		2.0		X				X			
C. spp.			1.0									
Endosporites globiformis	X	10.0	2.0	15.0		20.5		44.5		X		78.0
Laevigatosporites desmoinesensis	X	5.0	0.5			X			X	X	0.5	0.5
L. globosus		1.0	1.5						9.0			
L. medius		0.5										
L. ovalis	0.5	X	2.5	0.5	1.0	1.5	2.0	1.5	3.5	X	2.0	
L. vulgaris		X	0.5	X	X		0.5		X	X	0.5	
Punctatosporites minutus	22.0	3.5	25.5	0.5	6.5		40.5		9.0		11.0	
Latosporites minutus		X		0.5		0.5	0.5				1.5	
Spinosporites exiguus						X					0.5	
Tuberculatosporites robustus					X		X		X			
Thymospora pseudothiessenii	1.5		1.5		62.5		0.5		X		4.5	
Torispora securis	X								X			
Vestispora fenestrata	X		2.5			X	0.5			X	X	1.0
V. laevigata							X					
V. sp.				X								0.5
Florinites mediapudens	1.0	4.0	2.0	2.0		7.0	0.5	2.0				
F. millotti			0.5									
F. similis			X							X		0.5
F. triletus								2.5				
F. visendus								1.5				0.5
F. volans		0.5	X	0.5	X			1.0		X		
F. spp.			0.5					0.5				
Wilsonites circularis									X			
W. delicatus					2.0	X					X	
W. vesicatus	0.5	3.5	0.5	1.5	11.0	X	1.0			X		
Vesicaspora wilsonii		X								X	0.5	

*Coal below top bench of West Franklin Limestone at type section. Spore preservation not adequate for determining relative abundance.
†Formal systematics of taxa given in Appendix A.
§X = present but not in count.

TABLE 16. SPORE ANALYSIS OF UPPER DESMOINESIAN COAL BEDS MENTIONED IN THE TEXT; KANSAS: UNNAMED COAL BELOW SNI MILLS LIMESTONE MEMBER (MACERATION 2795), UNNAMED COAL ABOVE WORLAND LIMESTONE MEMBER (MACERATION 2805), DAWSON COAL (MACERATION 2796); OKLAHOMA: UNNAMED COAL BELOW NUYAKA CREEK LIMESTONE (MACERATION 2827), DAWSON COAL (MACERATIONS 2797A AND 2714); ILLINOIS: ROCK BRANCH (MACERATIONS 2596B-C), AND DEGRAFF (MACERATION 693E) COAL MEMBERS; PENNSYLVANIA: UPPER FREEPORT COAL (MACERATION 2816); OHIO: MAHONING COAL (MACERATIONS 2788 AND 3100N)

Spore taxa*	2795 (%)	2805 (%)	2796 (%)	2827 (%)	2797A (%)	2714 (%)	2596B (%)	2596C (%)	693E (%)	2816 (%)	2788 (%)	3100N (%)
Deltoidospora levis			1.0				X†					
D. priddyi							2.0					
D. subadnatoides				X								
D. spp.									0.5			
Punctatisporites flavus	X									X		
P. glaber					0.5							
P. minutus	1.0	1.0		2.0	0.5	5.5	2.0	2.0			0.5	0.5
P. spp.		0.5										
Calamospora breviradiata	X	2.0	0.5	1.0	1.0	0.5	0.5			1.5	1.0	1.0
C. hartungiana	0.5	X		X							X	
C. pedata		0.5			0.5							
C. spp.	0.5	1.5										0.5
Granulatisporites adnatoides	0.5						1.0				X	
G. granulatus			0.5									
G. livingstonensis	X	X										
G. minutus							0.5					
G. pallidus											0.5	
G. spp.		0.5		0.5					1.5			
Cyclogranisporites aureus	X											
C. obliquus						X			4.0	1.0	X	0.5
C. orbicularis		1.0	3.5	0.5	2.0	0.5	1.0	0.5	2.0	X	2.0	0.5
C. spp.	0.5	0.5	0.5		1.0							
Cadiospora magna		0.5					X					
Verrucosisporites sp.		X							X			
Kewaneesporites patulus		0.5										
Lophotriletes granoornatus										X		
L. rarispinosus					0.5							
Apiculatasporites latigranifer	0.5		0.5	X								
A. variusetosus							X					
Raistrickia abdita							0.5					
R. aculeata		X		X	0.5	0.5	X					
R. aculeolata						0.5					X	
R. crinita				X						X	X	
R. crocea	0.5						X					
Microreticulatisporites harrisonii			0.5									
M. nobilis								X			X	
Camptotriletes sp.										X		
Triquitrites additus			1.0									
T. bransonii	0.5		0.5	1.0	1.0		0.5		X	1.0	X	
T. exiguus			0.5		1.5		X					
T. pulvinatus										X		
T. spinosus			0.5	X		0.5	X	X		0.5	X	
T. spp.			0.5				X		0.5		0.5	
Mooreisporites inusitatus				X								

TABLE 16. SPORE ANALYSIS OF UPPER DESMOINESIAN COAL BEDS MENTIONED IN THE TEXT; KANSAS: UNNAMED COAL BELOW SNI MILLS LIMESTONE MEMBER (MACERATION 2795), UNNAMED COAL ABOVE WORLAND LIMESTONE MEMBER (MACERATION 2805), DAWSON COAL (MACERATION 2796); OKLAHOMA: UNNAMED COAL BELOW NUYAKA CREEK LIMESTONE (MACERATION 2827), DAWSON COAL (MACERATIONS 2797A AND 2714); ILLINOIS: ROCK BRANCH (MACERATIONS 2596B-C), AND DEGRAFF (MACERATION 693E) COAL MEMBERS; PENNSYLVANIA: UPPER FREEPORT COAL (MACERATION 2816); OHIO: MAHONING COAL (MACERATIONS 2788 AND 3100N) (continued)

Spore taxa*	2795 (%)	2805 (%)	2796 (%)	2827 (%)	2797A (%)	2714 (%)	2596B (%)	2596C (%)	693E (%)	2816 (%)	2788 (%)	3100N (%)
Crassispora kosankei	X	15.0		X	3.5	X	X		7.0		3.0	3.0
Granasporites medius	3.5	20.0	8.0	6.0	1.5	X	1.5	1.0	12.5	1.0	2.0	1.5
Lycospora brevijuga	1.0											
L. granulata	47.0	42.5	12.0	41.0	36.0	69.5	11.0	31.5	28.5	68.5	73.0	71.0
L. micropapillata		1.5				0.5				0.5	0.5	1.5
L. pusilla	1.0		1.0		0.5		0.5		14.5		1.0	0.5
L. rotunda		0.5	1.0								0.5	
L. subjuga											X	0.5
Cirratriradites annulatus		0.5								X	X	
C. annuliformis	X	X	1.5	X		X						
Endosporites globiformis		6.0	0.5	4.0	0.5			10.0				
E. plicatus		X		X		X						
Laevigatosporites desmoine-												
sensis	0.5	1.0	1.0				X	X		1.0	X	
L. globosus		0.5	1.0		1.5						2.0	4.5
L. ovalis	2.5	0.5	3.0	0.5	0.5	1.0	10.5	3.5	X	0.5		2.0
L. punctatus											0.5	
L. vulgaris		X	0.5	0.5			X		1.5		0.5	
Punctatosporites minutus	34.5	1.5	49.5	44.0	20.0	23.5	18.5	19.0	11.0	2.5	8.5	6.5
Latosporites minutus					0.5							0.5
Tuberculatosporites robustus							X					
Thymospora pseudothiessenii	2.5	9.0	1.5	2.5	22.0	1.0	46.5	41.5	1.0	22.0	4.0	5.0
Torispora securis			0.5									0.5
Vestispora fenestrata	X		0.5	X		X	X	X			X	
V. laevigata									X			
Florinites mediapudens	1.0	0.5	1.5	X	0.5	1.0	X	0.5				
F. similis		X							0.5			
F. triletus									1.0			
F. visendus								X				
F. volans				X								
Wilsonites circularis		X										
W. delicatus	2.0			X	0.5				2.0			
Pityosporites sp.						0.5	X		X			
Vesicaspora wilsonii								0.5	X			

*Formal systematics of taxa given in Appendix A.

†X = present but not in count.

TABLE 17. SPORE ANALYSIS OF LOWER MISSOURIAN COAL BEDS MENTIONED IN TEXT: OKLAHOMA "DUCK CREEK" COAL (MACERATION 2781), "CHECKERBOARD" COAL (MACERATION 2921); KANSAS: "HELPER" COAL (MACERATIONS 2808, 2831, AND 2892); MISSOURI: OVID COAL (MACERATION 772), UPPER LOCUST CREEK COAL (MACERATION 778), GRAIN VALLEY COAL (MACERATION 2779); ILLINOIS: ATHENSVILLE (MACERATION 2596D) AND CHAPEL (NO. 8) (MACERATION 573) COAL MEMBERS, AND THE UNNAMED COAL UNDERLYING CARLINSVILLE LIMESTONE MEMBER (MACERATION 577); OHIO: BRUSH CREEK COAL BED (MACERATIONS 3052 AND 2964)

Spore taxa*	2781 (%)	2921 (%)	2808 (%)	2831 (%)	2892 (%)	772 (%)	778 (%)	2779 (%)	2596D (%)	573 (%)	577 (%)	3052 (%)	2964 (%)
Deltoidospora grandis	X†			0.5				X					
D. levis	X		X				X						
D. priddyi			1.0	1.0	1.0								
D. subadnatoides	X						X						
D. sphaerotriangula			0.5										
D. spp.			0.5				X	0.5	X				X
Punctatisporites glaber			1.5	X				0.5					
P. minutus	21.0	1.0	5.5	2.0	1.5	72.0	64.0	5.0	17.5	66.5	55.0	74.0	76.5
P. spp.	2.0		0.5					1.0					
Calamospora breviradiata	3.0	17.0	4.0	5.0	8.5	6.5	3.5	12.0	3.5	1.0	4.5	X	1.5
C. hartungiana	2.0	0.5	1.0		0.5	2.0	1.0		X		7.0		X
C. straminea	2.0		0.5										
C. spp.	1.5			2.5		1.5	3.0	0.5		0.5			0.5
Granulatisporites granulatus		X										X	
G. livingstonensis			X										
G. minutus	1.0					0.5		3.0	5.0			0.5	0.5
G. spp.			0.5	0.5						1.5			
Cyclogranisporites aureus	X				X			X			1.5		
C. microgranus						X		X	0.5				X
C. minutus	0.5							X					
C. obliquus	11.5	23.0	28.5	14.5	17.0		0.5	2.0	2.0		3.5		1.0
C. orbicularis	2.0	15.0	11.5	6.0	21.5	8.5	11.0	40.5	40.0	13.0	10.5	3.0	7.0
C. spp.	0.5	1.0	0.5	1.0	1.5	0.5		6.0	9.0	2.0			2.5
Cadiospora magna	X			X		X	X	X					
Verrucosisporites grandiverrucosus	0.5						X						
V. microtuberosus		X										X	
V. henshawensis												X	
V. spp.					X			X		0.5		X	
Kewaneesporites patulus							X						
Lophotriletes commissuralis								X			X		
Apiculatasporites latigranifer	0.5	2.0		X	0.5	2.5	X	X			1.0	X	X
A. setulosus	X			X	0.5	X	0.5	0.5			X		X
A. variusetosus								0.5					
A. spp.	1.5												
Raistrickia aculeata			X				X						
R. carbondalensis							X	X					
R. crinita							X						X
R. crocea		0.5					X	X					
R. spp.								0.5					
Convolutispora sp.								0.5					
Microreticulatisporites nobilis	X						X	X					X
M. sulcatus	X				0.5		X	X	X				

TABLE 17. SPORE ANALYSIS OF LOWER MISSOURIAN COAL BEDS MENTIONED IN TEXT: OKLAHOMA "DUCK CREEK" COAL (MACERATION 2781), "CHECKERBOARD" COAL (MACERATION 2921); KANSAS: "HELPER" COAL (MACERATIONS 2808, 2831, AND 2892); MISSOURI: OVID COAL (MACERATION 772), UPPER LOCUST CREEK COAL (MACERATION 778), GRAIN VALLEY COAL (MACERATION 2779); ILLINOIS: ATHENSVILLE (MACERATION 2596D) AND CHAPEL (NO. 8) (MACERATION 573) COAL MEMBERS, AND THE UNNAMED COAL UNDERLYING CARLINSVILLE LIMESTONE MEMBER (MACERATION 577); OHIO: BRUSH CREEK COAL BED (MACERATIONS 3052 AND 2964) (continued)

Spore taxa*	2781 (%)	2921 (%)	2808 (%)	2831 (%)	2892 (%)	772 (%)	778 (%)	2779 (%)	2596D (%)	573 (%)	577 (%)	3052 (%)	2964 (%)
Triquitrites additus			1.0	1.5	3.5		X						
T. bransonii	4.5	2.5	3.0	22.5	7.0	1.5	1.0	1.5	X	0.5		X	X
T. crassus					1.0		X	3.0	X	X			X
T. exiguus	0.5				X								
T. protensus						X							
T. spinosus								X	1.0	X	0.5	X	X
T. subspinosus		X											
T. spp.	0.5												
Reinschospora magnifica												X	
R. triangularis									X				
Indospora stewartii												X	
Reticulatisporites muri- catus									X		X	X	
Crassispora kosankei							X				0.5	0.5	
Cirratriradites annuli- formis			X										
Endosporites globi- formis	27.5	30.5	27.5	32.5	18.5	2.0		0.5		7.5	8.5	X	X
E. plicatus	X							1.0		0.5		X	X
Laevigatosporites des- moinesensis		2.5	0.5	0.5	0.5							X	1.0
L. globosus		0.5	2.0	0.5							X		
L. medius								0.5	0.5				1.0
L. ovalis	2.0	0.5	3.5	1.5	7.0		3.0	10.5	3.0		0.5		
L. vulgaris					0.5			X	X	1.0			
Punctatosporites min- utus	3.0		3.0	1.5	1.5	5.0	1.0	6.0			0.5	1.5	0.5
Latosporites minutus	0.5		0.5		X		7.5			3.0	2.5	8.5	4.0
Spinosporites exiguus								5.0		1.5		10.0	1.5
Tuberculatosporites robustus		0.5							X			X	X
Thymospora obscura													X
T. pseudothiessenii					X								
Vestispora fenestrata	X	1.0		0.5	0.5				X				
Florinites mediapudens	3.0	2.0	0.5		6.0	0.5		1.0	2.0	1.0		1.5	2.5
F. similis													X
F. triletus											4.0		
F. visendus	3.5			X	X	X	X				X		
F. volans	2.0	X			X	0.5	X						
F. spp.	X												
Wilsonites delicatus	1.0	X	1.0		X			1.5			X		
W. vesicatus	2.0		1.5	8.5				7.0		4.5		0.5	X
Vesicaspora wilsonii	0.5				X		X						
Potonieisporites sp.						X							

*Formal systematics of taxa given in Appendix A.

†X = present but not in count.

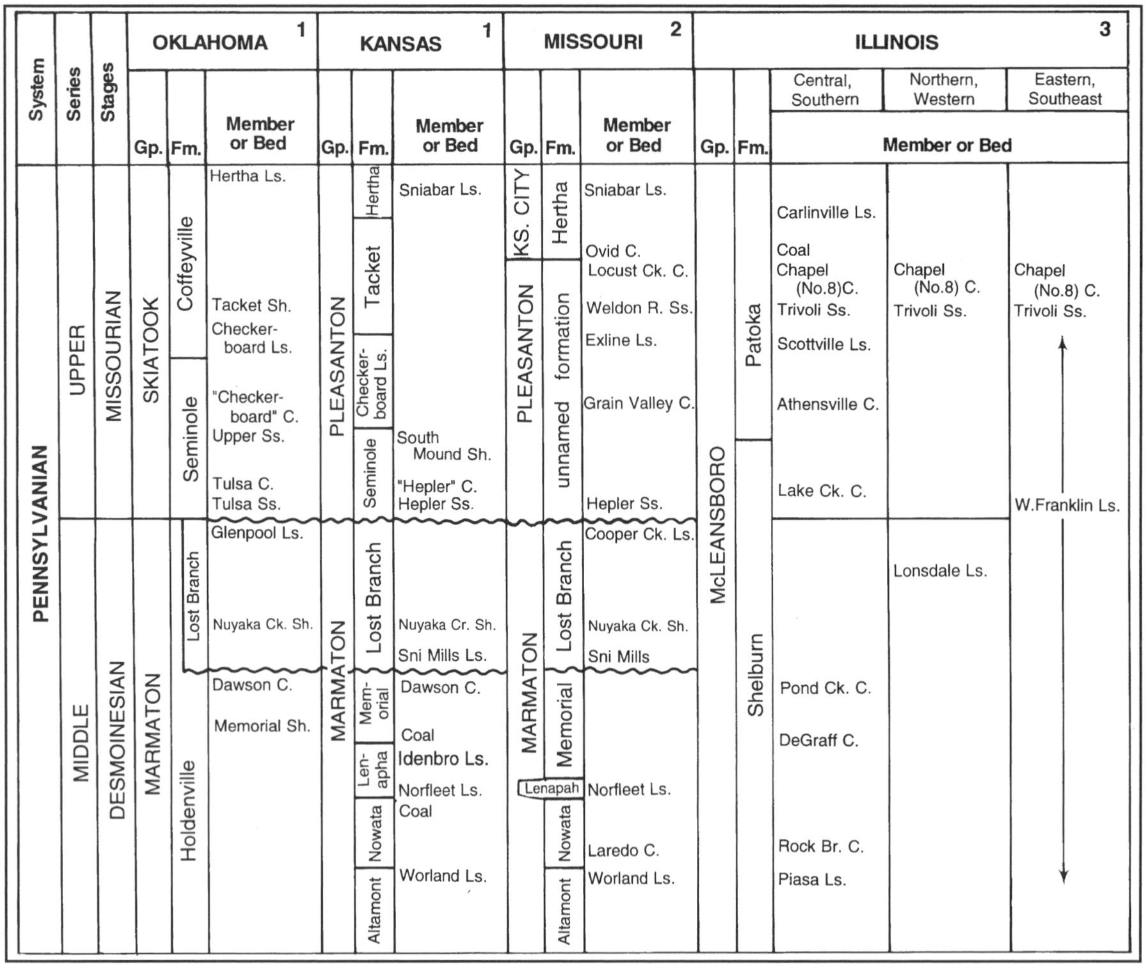

System	Series	Stages	OK Gp.	OK Fm.	OKLAHOMA Member or Bed	KS Gp.	KS Fm.	KANSAS Member or Bed	MO Gp.	MO Fm.	MISSOURI Member or Bed	IL Gp.	IL Fm.	ILLINOIS Central, Southern	Northern, Western	Eastern, Southeast
PENNSYLVANIAN	UPPER	MISSOURIAN	SKIATOOK	Coffeyville	Hertha Ls.	PLEASANTON	Hertha	Sniabar Ls.	KS. CITY	Hertha	Sniabar Ls.	McLEANSBORO	Patoka	Carlinville Ls.		
							Tacket			unnamed formation	Ovid C. Locust Ck. C.			Coal Chapel (No.8)C. Trivoli Ss.	Chapel (No.8) C. Trivoli Ss.	Chapel (No.8) C. Trivoli Ss.
				Seminole	Tacket Sh. Checker-board Ls.		Checker-board Ls.				Weldon R. Ss. Exline Ls.			Scottville Ls.		
					"Checker-board" C. Upper Ss.			South Mound Sh.			Grain Valley C.			Athensville C.		W.Franklin Ls.
					Tulsa C. Tulsa Ss.		Seminole	"Hepler" C. Hepler Ss.			Hepler Ss.			Lake Ck. C.		
	MIDDLE	DESMOINESIAN	MARMATON	Lost Branch	Glenpool Ls.	MARMATON	Lost Branch		MARMATON	Lost Branch	Cooper Ck. Ls.		Shelburn		Lonsdale Ls.	
					Nuyaka Ck. Sh.			Nuyaka Cr. Sh. Sni Mills Ls.			Nuyaka Ck. Sh. Sni Mills					
				Holdenville	Dawson C. Memorial Sh.		Mem-orial	Dawson C. Coal		Memorial				Pond Ck. C. DeGraff C.		
							Len-apha	Idenbro Ls.		Lenapah	Norfleet Ls.					
							Nowata	Norfleet Ls. Coal		Nowata	Laredo C.			Rock Br. C.		
							Altamont	Worland Ls.		Altamont	Worland Ls.			Piasa Ls.		

Figure 6. Correlation of upper Desmoinesian and lower Missourian strata in the Midcontinent. Adapted from (1) Heckel (1991), (2) Heckel (1991) and Howe (1982), and (3) Hopkins and Simon (1975).

overlying "Checkerboard" coal. In Tulsa County, sediments associated with the "Checkerboard" coal contain a great abundance of sigillarian compression fossils (Wilson, 1972, 1984), which was thought to be evidence of a major shift in kinds of floras just below the coal.

Pearson (1975), who studied the Tulsa and "Checkerboard" coals in Tulsa County, Oklahoma, did not observe *Thymospora pseudothiessenii, Granasporites,* or *Lycospora* in the Tulsa or "Checkerboard" coals. He noted that *Endosporites globiformis, Leschikisporis obliquus (Cyclogranisporites obliquus), Calamospora,* and *Florinites* increase in abundance from the Dawson coal to the middle Seminole (Tulsa) coal. He proposed placing the Desmoinesian-Missourian boundary at the base of the middle Seminole sandstone between the Dawson and Tulsa coals. Upshaw and Hedlund (1967) found minor percentages of *Lycospora* and *Thymospora* in the thick and rapidly deposited shale and clay in the Coffeyville Formation, which overlies the Seminole Formation in Oklahoma and may well have been redeposited from older strata. Neither were found in the coal in the Coffeyville Formation.

Samples of the Dawson coal from Tulsa (maceration 2714, Table 16) and Rogers (maceration 2715, Table 15) counties, Oklahoma, are dominated by *Lycospora. Punctatosporites minutus* is second in abundance. *Thymospora pseudothiessenii* is not common in either maceration. A sample (maceration 2797A, Table 16) of the Dawson coal from Nowata County has a spore assemblage composed of almost half *Lycospora. Punctatosporites minutus* is at about the same abundance as in the other two samples. Southward, in the Dawson coal in Okmulgee County, *L. granulata* and *P. minutus* greatly increase in abundance at the expense of *T. pseudothiessenii* (maceration 2827, Table 16). In the Tulsa coal of Missourian age, *Lycospora, Granasporites,* and *Thymospora* were not observed (maceration 2797B, Table 15), and *Punctatosporites minutus* is greatly reduced in abundance. *Cyclogranisporites* becomes the dominant genus, making up more than one-half of the assemblage, and *Endosporites* increases significantly in abundance. The "Duck Creek" coal (maceration 2781, Table 17) in Okmulgee County is probably equivalent to the Tulsa coal (Heckel, 1987, personal communication). In the "Duck Creek," *Cyclograni-*

sporites is less abundant, and *Endosporites* is more abundant than in the Tulsa coal.

The coal (maceration 2921) below the Checkerboard Limestone in the Tulsa region is dominated by *Endosporites globiformis, Cyclogranisporites obliquus, C. orbicularis,* and *Calamospora breviradiata* (Plate 4, Table 17). No specimens of *Lycospora, Thymospora pseudothiessenii, Punctatosporites,* or *Granasporites* were observed.

Heckel (1991) erected and described the Lost Branch Formation for strata between the top of the Holdenville Formation and Dawson coal and the base of the Seminole Formation in Oklahoma and equivalent strata in other parts of the Midcontinent. The paleontology of marine fossils demonstrates the presence of a major faunal change toward the top of the Lost Branch Formation (Heckel, 1984, 1986; Boardman et al., 1989b, 1990). The Glenpool Limestone bed at the top of the Lost Branch Formation contains a Desmoinesian fauna but a Missourian flora (Peppers, 1994). The Sni Mills Limestone Member at the base of the Lost Branch Formation underlies a thin shale that also contains a distinctive Desmoinesian conodont fauna. The same conodont fauna occurs in a thin shale in the Lonsdale Limestone Member of northern and western Illinois and in the lower part of the West Franklin Limestone Member in southeastern Illinois (Swade, personal communication, cited *in* Heckel, 1984). Bennison (cited *in* Heckel, 1984) reported the presence of a Missourian conodont fauna from the South Mound Shale Member, which overlies the "Hepler" coal.

In conclusion, a major palynological change occurs at the Desmoinesian-Missourian boundary in the western part of the Midcontinent between the Dawson and Tulsa coals or equivalent coals at about the base of the Pleasanton Group or Seminole Formation. The presence of occasional specimens of *Lycospora* in Missourian strata is not as significant in identifying the Desmoinesian-Missourian boundary outside the western part of the Midcontinent as are the great reduction in the abundance of *Lycospora* and other changes in spore assemblages that occur at the boundary.

Illinois basin. The Desmoinesian-Missourian boundary is correlated in this study with the base of the Lake Creek Coal Member in southern Illinois, with the top of the Lonsdale Limestone Member in northern and western Illinois, and with the top of the middle bench of the West Franklin Limestone Member in southeastern Illinois, western Indiana, and western Kentucky.

In the Illinois basin, the Desmoinesian-Missourian boundary had been correlated with the top of the West Franklin Limestone (Wanless, 1956; Kosanke et al., 1960; McKee and Crosby, 1975; Gray, 1979; Shaver et al., 1986), between the Athensville and Pond Creek Coal Members (Phillips et al., 1985), with the base of the Trivoli Sandstone Member (Dunbar and Henbest, 1942; Kosanke, 1950), with the top of the Trivoli Sandstone (Moore, 1944; Willman et al., 1967; Hopkins and Simon, 1975; and Atherton and Palmer, 1979), and with the overlying Chapel (No. 8) Coal Member above the Trivoli Sandstone (Rice et al., 1979b).

The stratigraphy in the interval between the Lonsdale Limestone and Trivoli Sandstone Members, which includes the Desmoinesian-Missourian boundary, has been somewhat uncertain; the coals are thin and only locally developed. Studies of marine limestone have been useful in correlating the Desmoinesian-Missourian in Illinois. Dunbar and Henbest (1942) found that a major change in fusulinid faunas occurs between the Lonsdale and Trivoli Limestones (now Cramer Limestone Member); therefore, they placed the Desmoinesian-Missourian boundary at the base of the Trivoli Sandstone, which overlies the Exline Limestone Member. The latter, which carries a Missourian fauna, was not studied by Dunbar and Henbest.

Kosanke (1947, 1950) stated that a major floral break occurs just below the Chapel Coal because *Lycospora* and *Thymospora pseudothiessenii* had not been observed in that coal. He also noted that *Lycospora* is dominant in the Scottville coal (now called Rock Branch Coal Member), is rare in the Upper Scottville coal (Athensville Coal Member), and is absent in the third Cutler Rider Coal (Lake Creek Coal Member). Kosanke (1950) interpreted the last observed occurrence of *Lycospora* and *Thymospora pseudothiessenii* as a more significant floral break than the change from abundance to scarcity of the two taxa. Kosanke (1947, 1950) therefore concluded that the floral change occurred at approximately the same position as the change in fusulinid faunas found by Dunbar and Henbest (1942), and placed the Desmoinesian-Missourian boundary at the base of the Trivoli Sandstone just below the Chapel Coal. Kosanke (1950) considered the first and second Cutler Rider coals (now DeGraff and Pond Creek Coal Members) in southern Illinois to be older than the Scottville and Upper Scottville coals of western and southwestern Illinois. The Rock Branch Coal was correlated with the Pond Creek Coal by Siever (cited *in* Wanless, 1956). Wanless (1956) and Kosanke et al. (1960) indicated that all of these coals are below the Lonsdale and West Franklin Limestones and correlated the Desmoinesian-Missourian boundary with the top of the limestones. Hopkins and Simon (1975) suggested that the Athensville Coal may correlate with the Lake Creek Coal and that the Pond Creek Coal may correlate with the Rock Branch Coal. Spore composition of these coals (Plate 4) does not support these correlations.

Andresen (1956) and Manos (1963) showed that the West Franklin Limestone in eastern and southeastern Illinois contains three limestone beds that probably correlate with the Piasa, Lonsdale, and Exline Limestone Members in other parts of Illinois (Fig. 6). Hopkins and Simon (1975) correlated the Desmoinesian-Missourian boundary with the top of the Trivoli Sandstone, which overlies the West Franklin Limestone. Thus, when the Desmoinesian-Missourian boundary was placed above the West Franklin Limestone, the Exline Limestone, which is Missourian in age, was erroneously included in the Desmoinesian.

A major change in spore populations, similar to that at the Desmoinesian-Missourian boundary in the Western Interior coal province, also occurred in the Illinois basin (Fig. 2) (Kosanke,

1950; Peppers and Phillips, 1972; Phillips et al., 1974; Phillips and Peppers, 1984; and Peppers, 1984, 1986). *Lycospora* diminished in abundance from about half of the spore assemblages in the youngest Desmoinesian coal to less than 1% in the overlying Missourian coals (Plate 4, Tables 15–17); *Thymospora pseudothiessenii* disappeared altogether. Occasional specimens of *Lycospora* and *Thymospora pseudothiessenii* occur, however, in Virgilian coals in the Illinois basin. *Granasporites* is abundant in the middle part of the Pennsylvanian but does not occur above the Middle-Upper Pennsylvanian boundary. *Laevigatosporites globosus* gradually disappeared, and *Punctatosporites minutus* almost disappeared at the boundary.

In the Desmoinesian Rock Branch Coal, *Thymospora pseudothiessenii* and *Lycospora* comprise about one-half and one-fourth respectively, of the spore assemblage. *Punctatosporites minutus* accounts for most of the remainder (Plate 4). *Lycospora* and other lycopsid spores (e.g., *Granasporites*, *Crassispora*, and *Endosporites*) increase in abundance to account for about three-fourths of the assemblage in the overlying DeGraff Coal. *Thymospora* and *Punctatosporites minutus* are less well represented. *Lycospora granulata* and *Punctatosporites minutus* greatly dominate the assemblage in the youngest Desmoinesian coal, the Pond Creek Coal. *Thymospora* is already reduced to less than 1% in the Pond Creek. In the oldest Missourian coal, the Lake Creek Coal, only a few specimens of *Lycospora* were observed, and *Granasporites* and *Thymospora* were not noted. *Endosporites* abruptly increases to make up almost one-half of the spore population; *Cyclogranisporites* is second in abundance (Table 15). *Lycospora, Granasporites,* and *Thymospora* also were not observed in the overlying Athensville Coal (Table 17). The fern spores *Cyclogranisporites* and *Punctatisporites minutus* increase greatly in abundance (Table 17). The spore population in the Chapel Coal and the unnamed coal below the Carlinville Limestone Member are quite similar to each other in that *Punctatisporites minutus* is dominant in both coals. This feature is common in many of the coals in the upper part of the Pennsylvanian.

Clendening (1974) and Kosanke (1988c) reported an abundance of several species of *Fabasporites* in Upper Pennsylvanian coals in the Appalachian region. The genus was described by Sullivan (1964) as lacking an aperture but having folds that resemble an aperture. In other aspects, *Fabasporites* appears to be identical to *Punctatisporites minutus*. Examination of spores from sporangia attached to specimens of *Asterotheca* (Pfefferkorn et al., 1971) demonstrates that some specimens clearly show an aperture, while others appear to be alete. Specimens of *Punctatisporites minutus* macerated from coal are almost always folded or broken; therefore, a suture is very difficult to demonstrate. I consider species of *Fabasporites* to be conspecific with *Punctatisporites minutus*.

Palynological analysis of a thin coal (maceration 831, Table 15) below the upper bench of the West Franklin Limestone at its type section in Posey County, Indiana, reveals that the coal is Missourian in age. No specimens of *Lycospora,* *Granasporites,* or *Thymospora* were observed; *Calamospora, Laevigatosporites, Cyclogranisporites,* and *Triquitrites* are the most abundant spores. The Pirtle Coal Member of Indiana is more than 60 ft (18.3 m) above the Danville Coal Member (VII) and a little below the West Franklin Limestone. At its type locality in Sullivan County, Indiana, the Pirtle Coal (maceration 2764, Table 15) has a typical Desmoinesian spore assemblage; more than one-half the population is made up of *Lycospora. Granasporites* and *Thymospora* are also common.

Appalachian coal region. The Desmoinesian-Missourian boundary (Middle-Upper Pennsylvanian boundary of the Midcontinent) is correlated herein with the top of the Mahoning coal or, if the coal is not present, the base of the Mahoning Sandstone in the lower part of the Conemaugh Group in Ohio and Pennsylvania. Because coal beds are not present in the upper part of the Charleston Formation and the lower part of the Conemaugh Formation in the proposed Pennsylvanian stratotype, the Desmoinesian-Missourian boundary has not been identified there.

The names Allegheny and Conemaugh have been widely used in the northern part of the Appalachian coal region since their introduction by Rogers (1840) and Platt (1875), respectively, in describing Pennsylvanian rocks in western Pennsylvania. The names have been used as formational names by the USGS, as group names by the Ohio and Pennsylvania Geological Surveys, and as series names by the Illinois Geological Survey (Wanless, 1956) and Indiana Geological Survey (Shaver et al., 1970). Campbell and Mendenhall (1896) applied the name Charleston Sandstone to strata between the Kanawha and Conemaugh Formations in central and southern West Virginia. The base of the Conemaugh is distinguished by the presence of variegated greenish gray and reddish shales that also commonly occur above the Mahoning Sandstone in Pennsylvania and Ohio. White (1908) and Krebs and Teets (1914) used the name Allegheny Series in their classifications of rocks in West Virginia. The name Charleston Sandstone was reinstated by Arkle (1969), and its rank was changed to group. Geologists with the USGS and others who are investigating the stratigraphic sections in the proposed Pennsylvanian System stratotype in Virginia and West Virginia, also use the name Charleston Sandstone. They designated its upper boundary as the top of the Middle Pennsylvanian.

According to Bradley (1956), the boundary between the Middle and Upper Pennsylvanian Series corresponds approximately to the boundary between the Allegheny and Conemaugh Formations of the Appalachian coal region and the boundary between the Desmoinesian and Missourian Series in the Midcontinent and other regions. Many stratigraphers working in the Appalachians use the Allegheny-Conemaugh boundary as the reference for the Middle and Upper Pennsylvanian boundary, but it is evident that the Desmoinesian-Missourian boundary is younger than the Allegheny-Conemaugh contact (Wanless, 1939; Moore, 1944; Cross and Schemel, 1952; Siever, cited *in* Wanless, 1956; Kosanke et al., 1960; Gray, 1979; Pfefferkorn and Gillespie, 1980; Phillips and Peppers, 1984; Phillips et al., 1985; Shaver et al., 1985, 1986). Palynological investigations support raising the USGS-proposed Middle-Upper Pennsylvanian boundary from the top of the Allegheny Formation to the top of the Mahoning coal or base of the Mahoning Sandstone Member in order to correspond with the Middle-Upper Pennsylvanian boundary in the Midcontinent and the Westphalian-Stephanian boundary.

Correlation of the upper boundary of the Charleston Sandstone in the proposed stratotype area with strata in western Pennsylvania is uncertain. Arndt (1979) matched the "Mahoning" Sandstone at the top of the Charleston Sandstone in West Virginia with the Mahoning Sandstone near the base of the Conemaugh Group in western Pennsylvania. Thus, the Middle-Upper Pennsylvanian boundary would be about equivalent to the Desmoinesian-Missourian boundary. However, Henry et al. (1979) indicated that the top of the Charleston Sandstone might be equivalent to the top of the Allegheny Group. By this interpretation, the "Mahoning" Sandstone in the Charleston Sandstone cannot be equivalent to the Mahoning Sandstone in the Conemaugh Formation unless the sandstone transgresses the Desmoinesian-Missourian boundary. Rice et al. (1979b) also correlated the top of the Allegheny Formation of Ohio with the top of the "Mahoning" Sandstone of West Virginia.

The Middle-Upper Pennsylvanian boundary cannot be correlated easily with strata outside the area of the proposed stratotype of the Pennsylvanian. The Upper Freeport coal, which marks the top of the Allegheny Formation in Ohio and Pennsylvania, is not present in the area of the proposed stratotype. The preservation of palynomorphs in the Conemaugh Formation of the proposed stratotype is poor compared with other areas in the northern Appalachians (Kosanke, 1988c). In addition, marine limestones are not common in the proposed stratotype (Englund et al., 1986); thus, good stratigraphic correlations using marine fossils are not possible.

Major changes in coal-swamp floras occurred in the lower part of the Monongahela Group in the northern part of the Appalachian coal region. Schemel (1957) reported that the Upper Freeport coal bed in West Virginia contains mostly *Lycospora granulata, Thymospora pseudothiessenii,* and *Laevigatosporites.* A spore assemblage that I examined from the Upper Freeport coal in Allegheny County, Pennsylvania, is greatly dominated by about two-thirds *L. granulata;* most of the remainder of the spore assemblage is made up of *T. pseudothiessenii* (Table 16, Plate 4).

The dominance of *Lycospora* in spore assemblages continues up into the Mahoning coal in the lower part of the Monongahela Group, but not into the Brush Creek coal (Phillips and Peppers, 1984). A sample of the Mahoning coal from Columbiana County, Ohio (maceration 2789, Table 15), is dominated by *Lycospora,* but *T. pseudothiessenii* and *Punctatosporites minutus* are well represented (Plate 4). Almost three-fourths of another sample (maceration 3100 N) from Columbiana County is made up of *Lycospora. Thymospora*

pseudothiessenii and *Granasporites medius* are also present. The Mahoning coal (Table 16) from Carroll County, Ohio, also yielded an assemblage of almost three-fourths *Lycospora. Thymospora pseudothiessenii* and *Punctatosporites minutus* are well represented. Only one bench of the Mahoning coal from a strip mine in Armstrong County, Pennsylvania, yielded spores, but the assemblage is similar to those in the coal in Ohio.

The composition of the palynomorph assemblage in the Mason coal, which occurs locally in Ohio, is strikingly different from that in the underlying Mahoning coal and is Missourian in age. A sample (Table 15) of the Mason coal collected in 1988, which is 15 to 20 ft (4.6 to 6.1 m) above the Mahoning coal in Columbiana County, contains 78% *Endosporites globiformis;* most of the remainder of the assemblage is made up of *Calamospora* and *Triquitrites bransonii. Lycospora, Thymospora pseudothiessenii, Granasporites,* and *Punctatosporites minutus* were not observed. Kosanke and Cecil (1989, 1992) also reported the occurrence of a major floral change between samples of the Mahoning and Mason coal beds at the locality from which I collected. It is remarkable that they recorded almost the same percentage (78.5%) of *Endosporites* in their sample of coal. The Mason Shale Member in Pennsylvania also contains an earliest Stephanian blattoid insect fauna (Durden, 1969, 1984a, 1984b).

Schemel (1957) reported less than 2% *Lycospora* in his samples of the Brush Creek coal, which overlies the Mason coal, and reported that rare specimens of the genus occur as high as the middle of the Conemaugh Formation. The Brush Creek coal contains rare specimens of *Thymospora pseudothiessenii,* but it contains an abundance of *Punctatisporites, Laevigatosporites, Endosporites globiformis, Florinites,* and *Punctatosporites minutus.* Schemel's abundant "Group 1" of *Laevigatosporites* is made up of small species, including *L. minutus* and *L. minimus,* but probably part of "Group 1" actually consists of *Punctatisporites minutus,* which he did not recognize.

Kosanke (1973) did not observe *Lycospora* or *Thymospora pseudothiessenii* in a coal sample thought to be the Brush Creek coal in Boyd County, Kentucky, but he thought that the poor preservation of spores may be responsible for their absence. *Punctatisporites minutus* accounts for about three-fourths of the spore assemblages in the samples (Table 17) of the Brush Creek coal from Ohio that I examined, and other fern spores account for most of the rest of the assemblages. No specimens of *Lycospora* or *Thymospora pseudothiessenii* were observed, but rare specimens of *T. obscura* occur. The age of the Brush Creek Limestone Member also has been identified as Missourian through the use of fusulinids (Smyth, 1974; Douglass, 1987), conodonts (Lane et al., 1971), and cephalopods (Unklesbay, 1954).

Kosanke (1988c, p. 3) stated that "palynological evidence from Ohio suggests that terrestrial plants underwent a major change just prior to the beginning of the *Triticites* range zone," which has been widely accepted as indicating the start of the Missourian. Thompson (1936) and Douglass (1987) reported that the oldest occurrence of *Triticites* is in the lower and upper Brush Creek Limestone Members in Ohio. As pointed out by Boardman et al. (1989b), however, recent studies indicate that *Triticites* may actually first appear a little above the Desmoinesian-Missourian boundary and that some of the early Missourian section is missing or condensed in the Appalachian region. Merrill (1964) used conodonts to correlate the Lower Brush Creek strata with the Dennis Formation, which is somewhat higher than the Desmoinesian-Missourian boundary in the northern part of the Midcontinent. The Lower Brush Creek was later correlated with the Swope Formation, which overlies the Dennis Formation, by Von Bitter and Merrill (1980) using conodonts and by Boardman et al. (1990) using ammonoids. The presence of *Thymospora obscura* in the Brush Creek coal also indicates that the Brush Creek may be a little younger than previously realized. In Illinois, except for an occasional specimen in the lower Desmoinesian, *T. obscura* first appears consistently in the New Haven Coal Member (Kosanke, 1950), which is several depositional cycles above the Desmoinesian-Missourian boundary. Douglass (1979) noted the presence of *Kansanella* sp. aff. *K. tenuis* in the Carthage Limestone in western Kentucky, which is similar to many fusulinids referred to as *Triticites ohioensis*. The Carthage Limestone overlies the New Haven Coal.

Westphalian-Stephanian boundary

Western Europe. The Westphalian-Stephanian boundary as originally defined at the 1935 International Congress of Carboniferous Stratigraphy and Geology is correlated with the Middle-Upper Pennsylvanian and Desmoinesian-Missourian boundaries.

The 1935 International Congress of Carboniferous Stratigraphy and Geology established the series classification of Carboniferous strata in western Europe (Jongmans and Gothan, 1937). Since then, the Subcommission on Carboniferous Stratigraphy has added numerous amendments. The Westphalian-Stephanian boundary was defined as being at the Holz Conglomerate in the Saar-Lorraine coal field, which is thought to mark a major lithostratigraphic and biostratigraphic break.

Cantabrian Stage. A recommendation was made and accepted by the Subcommission on Carboniferous Stratigraphy at the Seventh International Carboniferous Congress in 1971 that the Cantabrian should be recognized as a stage in the lowest part of the Stephanian (George and Wagner, 1972) and that the Holz Conglomerate be abandoned as the standard for the Westphalian-Stephanian boundary. The Saar-Lorraine section is used as a reference in this paper, however, because the palynology of the upper Westphalian and lower Stephanian has been better documented in this area than elsewhere in western Europe. Strata in Britain, northern France, and the Ruhr basin do not extend as high into the Westphalian D as at Saar-Lorraine. Bouroz et al. (1970) stated that the Stephanian directly overlying the Holz Conglomerate is not as old as the Stephanian in the Cevennes basin and several other basins in France. They

also stated that the boundaries of the Stephanian A, B, and C may not be correctly correlated. They proposed that the Cantabrian Stage and perhaps the early part of the Stephanian A are equivalent to the hiatus before the deposition of the Holz Conglomerate.

The Cantabrian was founded largely as a result of numerous studies by Wagner and his colleagues (e.g., Wagner, 1964, 1966, 1969; Wagner and Varker, 1971) on compression floras that occur in northern Spain but are not represented in the Saar-Lorraine basin. The stratotype of the Cantabrian Stage is in Palencia Province, northwestern Spain (Wagner et al., 1977; Wagner and Winkler Prins, 1985). Spain was part of a tectonic plate that was separated from the rest of Europe during the Upper Carboniferous (Scotese et al., 1979; Ziegler et al., 1979; Smith et al., 1981); chronostratigraphic correlation is therefore complicated because of differences in depositional facies.

Wagner and Winkler Prins (1985) stated that plant microfossils are poorly preserved at the stratotype of the Cantabrian, and no systematic effort has been yet made to recover microfossils. Therefore, the Cantabrian stratotype cannot be correlated palynologically to regions outside Spain. Wagner et al. (1970) reported on marine fossils, plant compression floras, and spores from eastern Asturia in northern Spain and concluded that the assemblages indicated a late Cantabrian age. They thought that the spore assemblages are younger than Westphalian D but not as young as the Stephanian in the Saar-Lorraine coal field. Châteauneuf (1973) correlated spore assemblages from central Asturia with the upper Westphalian D and lower Stephanian, although the assemblages are quite different from those in coals of comparable age in western Europe and North America. Diez and Cramer (1979) and Saenz de Santa Maria et al. (1985) concluded from their spore studies that the sections in the central Asturian coal field do not extend as high as the Stephanian. Laveine (1976) found that the fossil plant assemblages from the youngest strata in the central Asturian coal field would probably correspond to those between Tonsteins 60 and 40 of late Westphalian D age in the Saar-Lorraine coal field. Wagner and his colleagues have recently revised interpretations of their plant collections and now conclude that the Cantabrian is not represented in the Asturian Coal Basin (Winkler Prins, 1989, unpublished communication).

Alpern (1963), Alpern et al. (1960, 1967), Alpern and Liabeuf (1966, 1969), and Liabeuf and Alpern (1969) documented the palynological succession in the Saar-Lorraine coal field. They showed that a change in the composition of the spore floras occurs at the Holz Conglomerate. The most important change in plant megafossil assemblages in the Saar-Lorraine coal field also occurs at the Holz Conglomerate (Laveine, 1976). A second, less pronounced, break occurs at about Tonstein 60, which is at about the middle of the Westphalian D. Bode (discussion following Wagner and Winkler Prins, 1979) concluded that the Cantabrian corresponds to the upper part of the Westphalian D in the Saar basin.

It could be argued that the Cantabrian flora is not represented in the Saar-Lorraine region because of differences in depositional environments rather than the hiatus above the Holz Conglomerate. Remy (1975) thought that the Cantabrian flora is a deviation from the uniform Euramerican flora caused by local morphological and climatological differences. Wagner (discussion following Aizenverg et al., 1978, p. 169), however, stated that "there is no reason to assume that the Cantabrian flora would have a different composition because of facies consideration."

Illinois basin. The Westphalian-Stephanian boundary is herein correlated with the Desmoinesian-Missourian boundary in the Illinois basin and elsewhere in North America.

Correlation of the Westphalian-Stephanian boundary with the Desmoinesian-Missourian boundary is demonstrated by comparing miospore ranges and abundances in the Illinois basin with those in the Saar-Lorraine coal field of France; some early Stephanian strata, however, are missing below the Holz Conglomerate (Peppers, 1964; Plate 1). In the Saar-Lorraine coal field, *Lycospora* diminished markedly in abundance at the Westphalian-Stephanian boundary (Alpern, 1963; Alpern and Liabeuf, 1966; Alpern et al., 1967) but was still of significant occurrence in the Stephanian (Plate 4), especially in the limnic basins of central France (Liabeuf and Alpern, 1969, 1970; Liabeuf et al., 1967). Perhaps *Lycospora* was more abundant in these basins than it was in the paralic coal basins in North America because the swamps in France were not as affected by major marine regression as swamps were in North America. *Torispora* and *Thymospora* decreased greatly in abundance, but *Thymospora obscura* and *T. verrucosa (T. thiessenii)* became abundant in the upper Stephanian in the Jura and St. Etienne basins. *Punctatosporites granifer* declined in abundance a little above the Holz Conglomerate. The frequency of *Florinites, Laevigatosporites, Endosporites,* and the tree-fern spores *Punctatosporites rotundus* and *Cyclogranisporites obliquus* increased in abundance above the Westphalian-Stephanian boundary. *Granasporites medius* disappeared at the Desmoinesian-Missourian boundary in North America; however, Alpern (1959) reported that, although rare, it extended as high as the Permian in France. Bhardwaj (1957a) reported that *Lycospora* disappeared at the Holz Conglomerate and then resumed its range at the Stephanian B-C boundary in the Saar area. Several genera and species of megaspores declined significantly in abundance, and several disappeared at the Holz conglomerate.

Wagner (1984, p. 123) stated that the "Mazon Creek flora of Illinois, corresponding to the No. 2 Coal of the Carbondale Formation, correlates to the highest coal-measures in Britain and can be ascribed to the Westphalian D-Cantabrian transition." Remy (1975), using studies of plant compressions, also correlated the base of the Stephanian with the Colchester (No. 2) Coal. Wagner and Remy did not examine the flora of the Mazon Creek Shale Member, which overlies the Colchester Coal, but the flora is very well documented (Lesquereux, 1870; Noé, 1925; Janssen, 1939; Langford, 1958, 1963; Darrah, 1970; Peppers and Pfefferkorn, 1970; Phillips et al., 1973; Pfefferkorn, 1979; Horowitz, 1979).

According to palynological data (Peppers, 1984; this book), the Mazon Creek flora is about middle Westphalian D in age. Pfefferkorn (1979), who studied compression floras in the Illinois basin, concurred with this finding. As mentioned previously, the base of the epibole of *Thymospora* is at about the Dekoven Coal, which just underlies the Colchester Coal and Mazon Creek flora. Butterworth and Smith (1976) also pointed out that the base of the epibole of *Thymospora* is at Tonstein 60 at Saar-Lorraine and corresponds to the middle of the *Thymospora obscura* spore assemblage zone, which is Westphalian D in age. They stated that *Schopfites* is also recognized in the middle of the assemblage zone. In the Illinois basin, the genus is first recognized at the same level as the base of the epibole of *Thymospora*. Tonstein 60 marks the base of the *Lobatopteris vestita* floral zone of Wagner (1984).

Wagner (1984) placed the flora above the Herrin (No. 6) Coal (Plate 1) in Illinois, which was described by Gastaldo (1977), in the lower part of Wagner's *Odontopteris cantabrica* floral zone, which is above the *Lobatopteris vestita* zone and well within the Cantabrian Stage. The Herrin Coal is above the Colchester Coal and is late Desmoinesian in age, which is correlated with the late Westphalian D. Therefore, that part of the Cantabrian would be Westphalian D in age.

In conclusion, the Mazon Creek flora is correlated with Tonstein 60 of the Saar-Lorraine basin, which is middle Westphalian D in age. Because the flora has been considered transitional with the Cantabrian, strata between Tonstein 60 and the Holz Conglomerate in the Saar-Lorraine basin would have to be Cantabrian and upper Westphalian D (Plate 1). If all but the lower part of the Carbondale Formation is correlated with the Cantabrian, and the latter was deposited during the time represented by the hiatus associated with the Holz Conglomerate, a more transitional change from Desmoinesian to Missourian spore assemblages in Illinois would be expected. This is not the case, however, because a major change in floras occurs at the Desmoinesian-Missourian boundary. No significant disconformity occurs at the boundary in the Illinois basin, and only a minor disconformity occurs at the boundary in the Western Interior coal province. Sedimentation was also essentially continuous at the Middle-Upper Pennsylvanian boundary in the Appalachians (Arkle, 1969), and a major break in spore assemblages occurs there as well. Therefore, the abrupt change in Pennsylvanian peat floras cannot be explained as being caused entirely by a hiatus.

Compression floras should be reexamined to assess the existence of a Cantabrian flora in the United States. Palynologic correlation of the Cantabrian cannot be made with any confidence until the spore assemblages in its stratotype are described in detail.

Changes in palynological, floral, and faunal assemblages at the base of what has been considered Cantabrian in Illinois

(Colchester Coal) are not as pronounced as the changes that occur at the Desmoinesian-Missourian boundary. As Meyen stated (Bouroz et al., 1978, p. 77):

A strong provincialism of the Carboniferous fauna and flora and a certain amount of disagreement about stages in the evolution of various groups of organisms provoke certain difficulties in global correlation . . . A minor, but still quite important climatic change can be suggested at the Middle-Upper Carboniferous boundary. Focusing on the purely palaeontological criteria of major stratigraphic units one can easily lose the main purpose of an integrated stratigraphic scheme out of sight, a scheme, which should first of all constitute an effective instrument of correlation.

Appalachian coal region. The Westphalian-Stephanian boundary is correlated with just above the base of the Conemaugh Formation in the Appalachian coal region and with the top of the Mahoning coal or base of the Mahoning Sandstone in Ohio and Pennsylvania. It is a little younger than the Middle-Upper Pennsylvanian boundary of USGS usage.

Palynology of this distinctive boundary was discussed in the section dealing with the Desmoinesian-Missourian boundary. In addition, a middle Cantabrian to Stephanian A flora has been reported in the Narragansett basin in Rhode Island (Lyons, 1984). The Cantabrian Stage was determined using the earliest Stephanian floras in Spain, which are missing in France at the hiatus below the Holz Conglomerate (George and Wagner, 1972). The flora underlying the Cantabrian flora in the Narragansett basin was considered to be Westphalian D in age (Lyons, 1984) and was correlated with the Upper Freeport coal in Pennsylvania and Ohio. The Upper Freeport coal is almost the same age as the Danville Coal in Illinois. Thus, a Cantabrian flora first appears in Rhode Island near the top of the Westphalian and Desmoinesian. According to Wagner (1984), however, it begins in the Mazon Creek flora in Illinois, which is considerably older than the Danville Coal.

Donets basin. I am correlating the Westphalian-Stephanian boundary, as displayed at Saar-Lorraine, and the Desmoinesian-Missourian boundary with Limestone N_4 in the C_3^1 suite, which is a little above the Moscovian-Kasimovian boundary in the Donets basin.

Inosova et al. (1975, 1976) correlated the Westphalian-Stephanian boundary with Limestone N_4 near the base of the Kasimovian Stage. *Densosporites* and *Lycospora pusilla* declined abruptly in abundance to less than 3% in coal near Limestone N_4 (Plate 4), but the latter had a slight resurgence in the Autunian (Inosova et al., 1976). *Punctatosporites minutus, P. oculus,* and *Triquitrites* decreased in abundance, whereas *Calamospora, Cyclogranisporites, Cadiospora, Endosporites, Wilsonites,* and *Potonieisporites,* among other taxa, increased in abundance above the Westphalian-Stephanian boundary. *Vestispora laevigata* disappeared at the boundary, and *V. fenestrata* disappeared a little above the boundary. Seven species of spores or pollen that are characteristic of Permian assemblages appeared at the boundary.

Teteryuk (1976, 1982) correlated the Westphalian-Stephanian boundary with the Moscovian-Kasimovian and Desmoinesian-Missourian boundaries. Although Teteryuk (1974) correlated the Westphalian-Stephanian boundary with Limestone N_2, he stated that Limestone N_4 marks the base of spore zone SL_0 (Alpern and Liabeuf, 1969) at the Holz Conglomerate in the Saar-Lorraine Basin. Teteryuk (1974, p. 115) added that Limestone N_4 is at the boundary between the Middle and Upper Carboniferous and suggested that the floral changes at the boundary may have been caused by the onset of an arid climate in the Upper Carboniferous:

This phenomenon occurs in the palynological assemblages close to the boundary between the Westphalian and Stephanian in Saar-Lorraine (Alpern and Liabeuf, 1966), the basins of the Massif Central in France (Alpern et al., 1964; Alpern and Liabeuf, 1967), and of other countries in West Europe and North America (Hacquebard, Barss and Donaldson, 1960; Peppers, 1964) and is evidently characteristic of the entire paleofloristic region . . . This is where we see the end of the wide distribution of spores of the genus *Lycospora,* where *Torispora perverrucosa* Alpern and *Triquitrites mamosus* Bharadwaj appear, where *Vestispora fenestrata* (Kosanke and Brokaw) Wilson and Venkatachala disappears or are encountered very rarely, and where there is marked increase in the incidence of *Punctatosporites rotundus* Bharadwaj and *Punctatisporites obliquus* (?) Kosanke.

Teteryuk (1974) also thought that spore assemblages in the Limestone N_2–N_4 interval in the Donets basin are not completely represented at Saar-Lorraine because the strata are replaced by the Holz Conglomerate. Although he thought that the missing strata are equivalent to the Cantabrian strata, spore assemblages in the N_2–N_4 interval and Cantabrian rocks cannot be correlated because spore assemblages from the Cantabrian stratotype have not been systematically described. Novik and Fissunenko (1978) also stated that compression floras indicate that the Cantabrian Stage is between Limestones N_2 and N_4, at the base of the Kasimovian. However, Aizenverg (discussion following Aizenverg et al., 1978) stated that a Cantabrian flora is absent in the Donets basin because of its regional character.

Using two different interpretations of stratigraphic ranges of plant compressions, Fissunenko and Laveine (1984) concluded that the Westphalian-Stephanian boundary could correlate with either Limestone N_2 or Limestone 0_1. A number of Westphalian D and basal Cantabrian plant fossils disappeared below Limestone N_3 at a change in facies (Fissunenko, 1974). Perhaps the climatic change that occurred near the Westphalian-Stephanian boundary affected the peat swamp floras (as indicated by spore assemblages) slightly later than floras growing outside peat swamps (as indicated by compression fossils in clastic sediments) because of greater moisture in peat swamps.

Significant changes also occurred in the foraminifers, pelecypods, brachiopods, corals, and ostracode faunas in the interval between Limestones N_2–N_4 in the Donets basin (Teteryuk, 1974). Several species of Moscovian fusulinids disappeared at Limestone N_2, and Limestone N_3 marks the level of the first appearance of several species of *Fusulina* and *Fusulinella,* as well as an increase in *Protriticites* (Wagner and Higgins, 1979). Wagner and Varker (1971) and van Ginkel (1972) stated that

Cantabrian fusulinid faunas in northern Spain correlate with faunas in the highest Myachkovian beds. The base of the newly defined Stephanian would therefore be in the upper part of the Moscovian and what was formerly considered upper Westphalian D.

China. Palynological studies in China have not been mentioned yet in this book because, as noted by Owens et al. (1989), the palynomorph record in China is difficult to correlate precisely with that in the equatorial coal belt. Floral changes at the Westphalian-Stephanian boundary in China, however, are sufficiently pronounced to warrant a brief discussion. These changes were of a different kind and magnitude than those that took place in Euramerica, because the eastern Cathaysian and western Euramerican floral provinces began to differentiate at the Westphalian-Stephanian boundary (Li and Yao, 1982). As more palynological studies in China are completed, greater similarities between the spore assemblage in China and the Euramerican equatorial coal belt will probably become evident.

Gao (1984), who summarized data from palynological studies of Carboniferous strata in China, divided the succession into 12 spore assemblage zones, all of which are represented in the Ningwa basin in Shanxi Province. Assemblage zone X of late Westphalian age is characterized by an abundance of *Lycospora, Hadrohercos, Laevigatosporites vulgaris, Microreticulatisporites, Triquitrites, Torispora securis,* and *Ahrensisporites.*

The dominant genera in Zones XI and XII of Stephanian age are *Laevigatosporites, Torispora, Punctatosporites,* and *Thymospora. Lycospora granulata* disappeared, and *L. uber* (= *L. pellucida*), which is recorded as not being significant in numbers, is the only species of *Lycospora* that extended into zones XI and XII. *Reinschospora, Dictyotriletes bireticulatus,* and *Hadrohercos* disappeared at the end of the Westphalian, and *Ahrensisporites* and *Triquitrites* almost disappeared. *Densosporites annulatus, Punctatosporites granifer, P. punctatus, P. rotundus, Laevigatosporites desmoinesensis, Torispora laevigata, T. verrucosa, Thymospora obscura, Kosankeisporites (Illinites), Limitisporites,* and five species of *Florinites* began their ranges at the beginning of the Stephanian. *Thymospora pseudothiessenii* occurs only in assemblage zone XII, but is rare. In Illinois, by contrast, *Densosporites annulatus* extends only as high as the late Westphalian C (Peppers, 1984), and *Thymospora pseudothiessenii* first appeared near the beginning of the Westphalian D and essentially disappeared at the Westphalian-Stephanian boundary. In Illinois, *Illinites (Kosankeisporites)* began its range in the late Westphalian B, *Punctatosporites punctatus* began in the middle Westphalian C, and *Triquitrites* was abundant in the Stephanian.

A Westphalian spore assemblage from northern Shanxi Province indicates an abundance of marattialean fern spores with only a few percent of the lycopod spore, *Densosporites,* but the presence of *Lycospora* was not reported (Ouyang and Li, 1980). This is quite different from the Westphalian assemblages that contain an abundance of *Lycospora* (Gao, 1984).

Li and Zhang (1983) found similar changes in spore assemblages at the Westphalian-Stephanian boundary in north China. In addition to the spore taxa discussed by Gao (1984), they reported that *Reticulatisporites muricatus, Dictyotriletes bireticulatus, Anapiculatisporites minor,* and *Cirratriradites saturni* disappeared at the Westphalian-Stephanian boundary. *Tripartites* extends to the top of the Upper Carboniferous in northern China but is limited to the Mississippian in Illinois.

Missourian-Virgilian Stage boundary

Western Interior coal province. Moore (1932) proposed the Virgilian Series (considered a stage in this book) for strata exposed near Virgil, Kansas, to separate the upper part of the Pennsylvanian from the Missourian on the basis of an unconformity and important faunal changes. The boundary was placed at the base of the Tonganoxie Sandstone, and the placement was accepted by a conference of representatives from several Midcontinent states (Moore, 1948). Because paleontological criteria for identifying the Missourian-Virgilian boundary were not included in the original definition of the Virgilian, precise correlation of the boundary is uncertain. The history of these correlations was discussed by Boardman et al. (1989b). Boardman et al. (1989a) recommended placing the Missourian-Virgilian boundary at the base of the Haskell Limestone Member in the Lawrence Formation in Kansas on the basis of the first appearance of certain ammonoids and conodonts just above the limestone. Marine limestones and light-colored shale are common, and coal beds and carbonaceous shales are rare in the Virgilian in the western part of the Midcontinent; therefore, palynology may not be very useful in defining the boundary and correlating it with strata in other regions.

Illinois basin. The Missourian-Virgilian boundary is correlated with the top of the Calhoun Coal Member in Illinois or, if the coal is not present, the base of the Omega Limestone Member.

The Virgilian Stage includes the upper half of the Mattoon Formation in the Illinois basin, which extends to the top of the Pennsylvanian. The stratigraphic relationships of some of the coals in the Virgilian and upper Missourian in the Illinois basin are not well understood, but the Missourian-Virgilian boundary was placed a few meters below an unnamed coal below the Shumway Limestone Member by Kosanke et al. (1960), Willman et al. (1967), and Hopkins and Simon (1975).

The Missourian-Virgilian boundary divides the ranges of the more primitive *Triticites* subgenus *Kansanella* and more advanced forms of *Triticites.* Dunbar and Henbest (1942) used fusulinids to correlate the boundary with a position between the Omega Limestone and the overlying Shumway Limestone Member. The boundary was placed lower, a little above the Livingston Limestone Member, by Cooper (1946) on the basis of ostracode faunas. Langenheim and Scheihing (1983), Scheihing and Langenheim (1985), and Weibel (1986), using mostly brachiopods, correlated the Bogota Limestone Member, which overlies the Shumway, with the middle of the Shawnee Group of Kansas, which overlies the Lawrence Formation. Heckel (1990, personal communication) correlated by use of conodonts the base of the Haskell Limestone in Kansas with the base of the Omega Limestone that overlies the Calhoon Coal in the Effingham-Newton area of Illinois.

Plant compressions are not common in Virgilian rocks of Illinois, and because most of the Virgilian coals are thin, of little economic importance, and rarely crop out, their palynology has not been studied in detail. Kosanke (1950) reported on the spores in several coals below and above the Missourian-Virgilian boundary. He found a small number of taxa in the Shelbyville Coal Member, which is a little below the boundary. *Endosporites*, *Laevigatosporites*, and *Punctatosporites* are the most abundant genera. *Endosporites* doubles in abundance in the Trowbridge Coal Member above the boundary, and taxa below and above the boundary are long ranging. The ranges and general abundance of the most common palynomorphs in the Mattoon Formation were given by Peppers (1984). Because ranges of other less-common Upper Pennsylvanian spores in the Midcontinent remain to be delineated, precise palynological correlations are not possible at this time.

The youngest coals found thus far in the Illinois basin are from Kentucky Geological Survey core Gil 30 drilled in a graben in the Rough Creek fault system in Union County, Kentucky. Several coal beds at a depth of 390 to 540 ft (118.9 to 165 m) contain spore assemblages characterized by the abundance of *Thymospora thiessenii* (Hower et al., 1983; Peppers, unpublished data). *Thymospora thiessenii* is also characteristic of the Pittsburgh coal bed (Cross, 1952; Clendening and Gillespie, 1964) and several other coal beds near the base of the Conemaugh Formation in the northern part of the Appalachian region. Douglass (1987) reported the fusulinid *Triticites beardi* from limestone at about 194 ft (59 m) that "represents a developmental stage similar to forms described from rocks of Early Permian (Wolfcampian) age." The *Thymospora thiessenii* spore assemblage zone also occurs in several coals in a core drilled in Gallatin County, Illinois (Peppers, 1984).

SUMMARY

Palynological data confirm many previously published correlations of major Pennsylvanian and Middle and Upper Carboniferous chronostratigraphic boundaries, but I have proposed several new correlations. The correlations were made on the basis of published reports and new information obtained from recent studies. In particular, correlations between the Illinois basin and Appalachian coal region were aided by palynological investigation of coal beds in eastern Kentucky and Tennessee. Palynological study of coal beds above and below the Desmoinesian-Missourian boundary in the Western Interior coal province also refined the correlation of that boundary with strata in the Illinois basin, Appalachian coal region, and Europe.

Correlations of major chronostratigraphic boundaries in the

Pennsylvanian and Middle and Upper Carboniferous are summarized in the following numbered paragraphs.

1. The Namurian-Westphalian boundary is not well documented in the Illinois basin, but I have tentatively correlated it with the top of the Battery Rock Sandstone Member in Illinois, just above the "Hindostan Whetstone Beds" in the lower part of the Mansfield Formation in Indiana, and the top of the Kyrock Sandstone in western Kentucky.

2. The boundary between the Lower and Middle Pennsylvanian Series, as proposed by the USGS, has been designated as the top of the New River Formation in the area of the proposed stratotype for the Pennsylvanian. It is older than the Lower-Middle Pennsylvanian boundary as used in the Midcontinent region and is correlated with the base of the Dye Shale in Arkansas. I am correlating the boundary with the base of the Pounds Sandstone Member just above the Gentry Coal Bed in Illinois. In Indiana, the boundary is in the Mansfield Formation about midway between the "Hindostan Whetstone Beds" and the St. Meinrad Coal Member; in western Kentucky, it is in the Caseyville Formation at the base of the Bee Springs Sandstone Member. The Gentry Coal (formerly Battery Rock Coal Member) in Illinois is older than the "Battery Rock coal bed" of western Kentucky. The Lower-Middle Pennsylvanian Series boundary of the USGS is approximately equivalent to Limestone H_1 (middle Bashkirian) in the Donets basin and is about middle Westphalian A in age.

3. The boundaries between the Westphalian A and B and Morrowan and Atokan Stages (Lower-Middle Pennsylvanian Series boundary in the Midcontinent) are correlative and occur just above the base of the Tradewater Formation, at the top of the Reynoldsburg Coal Bed in Illinois and a little below the St. Meinrad Coal Member in Indiana and Bell coal of western Kentucky. They correlate approximately with the top of the Lower War Eagle coal bed in West Virginia, the top of the Gray Hawk coal bed in eastern Kentucky, and the top of the Hooper coal bed in Tennessee. The boundaries are approximately equivalent to Limestone H_4 in the middle of the C_2^3 suite (upper Bashkirian) in the Donets basin.

4. The Bashkirian-Moscovian boundary of the Donets basin is herein considered late Westphalian B in age. It is correlated with a position a little below the middle of the bottom half of the Tradewater Formation in Illinois and in the lower part of the Mansfield Formation in Indiana. It is early Atokan in age. It is at the top of the Upper Elkhorn coal bed in eastern Kentucky, the Joyner coal bed in Tennessee, and the top of the Cedar Grove coal bed in West Virginia.

5. I am correlating the Westphalian B-C boundary with approximately the middle of the lower half of the Tradewater Formation in Illinois, and a little below the Smith coal bed in western Kentucky and Blue Creek Coal Member in Indiana. It is equivalent to the middle of the Kilbourn Formation in Iowa. The boundary occurs at the base of the Fire Clay coal in eastern Kentucky, the Windrock coal bed in eastern Tennessee, and base of

the Hernshaw coal bed in West Virginia. The Westphalian B-C boundary is approximately equivalent to Limestone K_7 near the top of the C_2^5 suite in the Donets basin.

6. The Atokan-Desmoinesian boundary, which is at the base of the Hartshorne Formation in the Midcontinent, is either just above or just below the Seville Limestone Member in Illinois, the Curlew Limestone Member in western Kentucky, and the Perth Limestone Member in Indiana, depending on various interpretations of microfaunal evidence. Palynological data indicate that the boundary most likely is at the top of the limestones. If the Atokan-Desmoinesian boundary is raised to the top of the limestones, which I propose, it would coincide with the Westphalian C-D boundary. The Atokan-Desmoinesian boundary is correlated with the base of the Princess No. 5B coal bed in eastern Kentucky, the base of the Upper Mercer coal bed in Ohio, and the base of the Upper No. 5 Block coal bed in West Virginia. It is equivalent to Limestone M_3 in the lower part of the C_2^7 suite in the Donets basin.

7. I conclude that the Middle-Upper Pennsylvanian boundary in the Appalachian coal region is approximately equivalent to the Moscovian-Kasimovian boundary in the Donets basin. The boundary between the Allegheny and Conemaugh Formations in the northern part of the Appalachian coal region has been used as the reference for the Middle-Upper Pennsylvanian boundary (Bradley, 1956). Palynological and paleontological evidence demonstrate that the Allegheny-Conemaugh boundary is older than the Desmoinesian-Missourian boundary. If the Desmoinesian-Missourian boundary is used as the basis for dividing the Middle and Upper Pennsylvanian Series as it is in the Midcontinent, the latter boundary would be a more easily traceable and chronostratigraphically useful boundary. The Middle-Upper Pennsylvanian boundary proposed by the USGS correlates with the top of the Danville (No. 7) Coal Member in Illinois and Indiana.

8. The Desmoinesian-Missourian Stage boundary (Middle-Upper Pennsylvanian Series boundary in the Midcontinent) and corresponding Westphalian-Stephanian boundary are characterized by a distinct change in composition of spore and faunal assemblages. Stratigraphers in the Midcontinent have recognized the base of the Pleasanton Formation in Missouri as the standard for the base of the Missourian Series. The Desmoinesian-Missourian boundary in the northern part of the Appalachian coal region is correlated with the top of the Mahoning coal bed or, if the coal is not present, with the base of the Mahoning Sandstone in the northern part of the Appalachian coal region. In Illinois, I place the Desmoinesian-Missourian boundary at the base of the Lake Creek Coal Member or at the top of the Lonsdale Limestone Member if the coal is absent. The boundary is correlated with the top of the middle bench of the West Franklin Limestone Member in Indiana and western Kentucky. Limestone N_4 in the C_3^1 suite (lower Kasimovian) probably marks the Westphalian-Stephanian boundary in the Donets basin. Palynological evidence does not support plant compression studies that indicate

placement of the base of the Cantabrian Stage (base of the Stephanian) at the interval of the Mazon Creek flora in Illinois (middle Westphalian D).

9. The Missourian-Virgilian boundary is herein tentatively correlated, on the basis of ammonoid and conodont studies, with the base of the Haskell Limestone Member in Kansas and the top of the Calhoun Coal Member or, if the coal is not present, with the base on the Omega Limestone Member.

ACKNOWLEDGMENTS

I am indebted to Rodney Norby, Philip DeMaris, and John Nelson (ISGS), Tom L. Phillips (University of Illinois), Robert Ravn (Aeon Biostratigraphic Services, Anchorage, Alaska), and C. R. Klug (University of Iowa) for reviewing the manuscript and Philip Heckel (University of Iowa) for reviewing parts of the manuscript. Robert M. Kosanke (USGS) reviewed an earlier version of the paper. The following individuals provided coal samples that were critical to some of the correlations: Blaine Cecil and Thomas Kehn (USGS); Horace R. Collins (Ohio Geological Survey); Allen Williamson, John Beard, James C. Cobb, James C. Currens, and John Chesnut (Kentucky Geological Survey); James Hower (Center for Applied Energy Research, Kentucky); Philip H. Heckel (University of Iowa); Allan Bennison (Tulsa, Oklahoma); Patrick Sutherland and Mark Dennen (University of Oklahoma); Lindgren Lin Chyi (University of Akron); Michael Caudill (University of Tennessee); Royal Mapes (Ohio University); and Laurence Nuelle (Division of Geology and Land Survey, Missouri). The illustrations were prepared by Vicky Reinhart (ISGS).

APPENDIX A. FORMAL SYSTEMATICS OF SPECIES OF SPORES DISCUSSED IN TEXT AND LISTED IN THE TABLES

Deltoidospora adnata (Kosanke) n. comb.
—1950 *Granulatisporites adnatus* Kosanke, p. 20, plate 3, Fig. 9.
D. grandis (Kosanke) n. comb.
—1950 *Granulatisporites grandis* Kosanke, p. 21, plate 1, Fig. 10.
D. levis (Kosanke) Ravn 1986
D. priddyi (Berry) McGregor 1973
D. pseudolevis (Peppers) Ravn 1986
D. sphaerotriangula (Loose) Ravn 1986
D. subadnatoides (Bhardwaj) Ravn 1986
Punctatisporites cf. *edgarensis* Peppers 1970
P. flavus (Kosanke) Potonié and Kremp 1955
P. glaber (Naumova) Playford 1962
P. incomptus Felix and Burbridge 1967
P. irrasus Hacquebard 1957
P. minutus (Kosanke) Peppers 1964
P. obesus (Loose) Potonié and Kremp 1955
Calamospora breviradiata Kosanke 1950
C. hartungiana Schopf (in Schopf, Wilson, and Bentall 1944)
C. liquida Kosanke 1950

C. mutabilis (Loose) Schopf, Wilson, and Bentall 1944

C. pedata Kosanke 1950

C. straminea Wilson & Kosanke 1944

Granulatisporites adnatoides (Potonié and Kremp) Smith and Butterworth 1967

G. granularis Kosanke 1950

G. granulatus Ibrahim 1933

G. livingstonensis Peppers 1970

G. microgranifer Ibrahim 1933

G. minutus Potonié and Kremp 1955

G. pallidus Kosanke 1950

G. piroformis Loose 1934

G. tuberculatus Hoffmeister, Staplin, and Malloy 1955

G. verrucosus (Wilson and Coe) Schopf, Wilson, and Bentall 1944

Cyclogranisporites aureus (Loose) Potonié and Kremp 1955

C. leopoldi (Kremp) Potonié and Kremp 1955

C. microgranus Bhardwaj 1957 (Bhardwaj, 1957a)

C. minutus Bhardwaj 1957 (Bhardwaj, 1957a)

C. obliquus (Kosanke) Upshaw and Hedlund 1967

C. orbicularis (Kosanke) Potonié and Kremp 1955

C. staplinii (Peppers) Peppers 1970

Cadiospora magna Kosanke 1950

Sinuspores sinuatus (Artüz) Ravn 1986

Verrucosisporites donarii Potonié and Kremp 1955

V. grandiverrucosus (Kosanke) Smith and others 1964

V. henshawensis (Peppers) Ravn 1986

V. microtuberosus (Loose) Smith and Butterworth 1967

V. morulatus (Knox) Smith and Butterworth 1967

V. sifati (Ibrahim) Smith and Butterworth 1967

V. verrucosus (Ibrahim) Ibrahim 1933

Kewaneesporites patulus (Peppers) Peppers 1970

Lophotriletes commissuralis (Kosanke) Potonié and Kremp 1955

L. copiosus Peppers 1970

L. gibbosus (Ibrahim) Potonié and Kremp 1955

L. granoornatus Artüz 1957

L. ibrahimii (Peppers) Pi-Radondy and Doubinger 1968

L. insignitus (Ibrahim) Potonié and Kremp 1955

L. microsaetosus (Loose) Potonié and Kremp 1955

L. mosaicus Potonié and Kremp 1955

L. pseudaculeatus Potonié and Kremp 1955

L. rarispinosus Peppers 1970

Waltzispora prisca (Kosanke) Sullivan 1964

Anapiculatisporites baccatus (Hoffmeister, Staplin and Malloy) Ravn 1986

A. grundensis Peppers 1970

A. minor (Butterworth and Williams) Smith and Butterworth 1967

A. spinosus (Kosanke) Potonié and Kremp 1955

Procoronaspora dumosa (Staplin) Smith and Butterworth 1967

Pustulatisporites crenatus Guennel 1958

P. grumosus (Ibrahim) n. comb.

—1955 *Apiculatisporites grumosus* (Ibrahim) Potonié and Kremp, p. 79, plate A, Figs. 242 and 243

Apiculatasporites latigranifer (Loose) Ravn 1986

A. setulosus (Kosanke) Ravn 1986

A. spinososaetosus (Loose) Ravn 1986

A. spinulistratus (Loose) Ibrahim 1933

A. variocorneus (Sullivan) Ravn 1986

A. variusetosus (Peppers) Ravn 1986

Planisporites granifer (Ibrahim) Knox 1950

Pilosisporites aculeolatus (Kosanke) Ravn 1986

P. triquetrus Smith and Butterworth 1967

P. williamsii (Knox) Ravn 1986

Raistrickia abdita (Loose) Schopf, Wilson, and Bentall 1944

R. aculeata Kosanke 1950

R. aculeolata Wilson and Kosanke 1944

R. breveminens Peppers 1970

R. carbondalensis Peppers 1970

R. crinita Kosanke 1950

R. crocea Kosanke 1950

R. fulva Artüz 1957

R. prisca Kosanke 1950

R. saetosa (Loose) Schopf, Wilson, and Bentall 1944

R. subcrinita Peppers 1970

R. superba (Ibrahim) Schopf, Wilson, and Bentall 1944

Spackmanites habibii Ravn 1986

Convolutispora florida Hoffmeister, Staplin, and Malloy 1955

C. fromensis Balme and Hassell 1962

C. mellita Hoffmeister, Staplin, and Malloy 1955

Microreticulatisporites concavus Butterworth and Williams 1958

M. lunatus (Knox) Knox 1950

M. harrisonii Peppers 1970

M. nobilis (Wicher) Knox 1950

M. sulcatus (Wilson and Kosanke) Smith and Butterworth 1967

Secarisporites remotus Neves 1961

Dictyotriletes bireticulatus (Ibrahim) Potonié and Kremp 1954

Camptotriletes bucculentus (Loose) Potonié and Kremp 1955

C. confertus (Ravn) Ravn 1986

Ahrensisporites guerickei (Horst) Potonié and Kremp 1954

A. guerickei var. *ornatus* Neves 1961

Triquitrites additus Wilson and Hoffmeister 1956

T. bransonii Wilson and Hoffmeister 1956

T. crassus Kosanke 1950

T. exiguus Wilson and Kosanke 1944

T. mamosus Bhardwaj 1957 (Bhardwaj, 1957a)

T. minutus Alpern 1958

T. protensus Kosanke 1950

T. pulvinatus Kosanke 1950

T. sculptilis (Balme) Smith and Butterworth 1967

T. spinosus Kosanke 1943

T. subspinosus Peppers 1970

T. tribullatus (Ibrahim) Schopf, Wilson and Bentall 1944

Zosterosporites triangularis Kosanke 1973

Mooreisporites bellus Neves 1961

M. fustus Neves 1958

M. inusitatus (Kosanke) Neves 1958

M. terjugas (Ishchenko) Teteryuk 1976

M. trigallerus Neves 1961

Reinschospora magnifica Kosanke 1950

R. triangularis Kosanke 1950

Indospora stewartii Peppers 1964

Knoxisporites seniradiatus Neves 1961

K. stephanephorus Love 1960

K. triradiatus Hoffmeister, Staplin, and Malloy 1955

Reticulatisporites mediareticulatus Ibrahim 1933

R. muricatus Kosanke 1950

R. polygonalis (Ibrahim) Smith and Butterworth 1967

R. reticulatus (Ibrahim) Ibrahim 1933
R. reticulocingulum (Loose) Loose 1934
R. splendens Kosanke 1950
Reticulitriletes falsus (Potonié and Kremp) Ravn 1986
Savitrisporites asperatus Sullivan 1964
S. concavus Marshall and Smith 1965
S. nux (Butterworth and Williams) Smith and Butterworth 1967
Grumosisporites varioreticulatus (Neves) Smith and Butterworth 1967
Crassispora kosankei (Potonié and Kremp) Smith and Butterworth 1967
Granasporites medius (Dybova and Jachowicz) Ravn, Butterworth, Phillips, and Peppers 1986
Murospora kosankei Somers 1952
Simozonotriletes intortus (Waltz) Potonié and Kremp 1954
Densosporites annulatus (Loose) Smith and Butterworth 1967
D. duriti Potonié and Kremp 1956
D. glandulosus Kosanke 1950
D. irregularis Hacquebard and Barss 1957
D. ruhus Kosanke 1950
D. sinuosus Kosanke 1950
D. sphaerotriangularis Kosanke 1950
D. spinifer Hoffmeister, Staplin, and Malloy 1955
D. triangularis Kosanke 1950
D. variabilis (Waltz) Potonié and Kremp 1956
Lycospora brevijuga Kosanke 1950
L. granulata Kosanke 1950
L. micropapillata (Wilson and Coe) Schopf, Wilson, and Bentall 1944
L. noctuina Butterworth and Williams 1958
L. orbicula (Potonié and Kremp) Smith and Butterworth 1967
L. pellucida (Wicher) Schopf, Wilson, and Bentall 1944
L. pusilla (Ibrahim) Schopf, Wilson, and Bentall 1944
L. rotunda Bhardwaj 1957 (Bhardwaj, 1957a)
L. subjuga Bhardwaj 1957 (Bhardwaj, 1957b)
Cristatisporites connexus Potonié and Kremp 1955
C. indignabundus (Loose) Staplin and Jansonius 1964
Paleospora fragila Habib 1966
Cirratriradites annulatus Kosanke and Brokaw (in Kosanke 1950)
C. annuliformis Kosanke and Brokaw (in Kosanke 1950)
C. maculatus Wilson and Coe 1940
C. rarus (Ibrahim) Schopf, Wilson, and Bentall 1944
C. reticulatus Ravn 1979
C. saturnii (Ibrahim) Schopf, Wilson, and Bentall 1944
Cingulizonates loricatus (Loose) Butterworth and Smith (*in* Butterworth and others) 1964
Radiizonates aligerens (Knox) Staplin and Jansonius 1964
R. difformis (Kosanke) Staplin and Jansonius 1964
R. rotatus (Kosanke) Staplin and Jansonius 1964
R. striatus (Knox) Staplin and Jansonius 1964
Endosporites globiformis (Ibrahim) Schopf, Wilson, and Bentall 1944
E. plicatus Kosanke 1950
E. staplinii Gupta and Boozer 1969
E. zonalis (Loose) Knox 1950
Schulzospora rara Kosanke 1950
Alatisporites hexalatus Kosanke 1950
A. hoffmeisterii Morgan 1955
A. pustulatus (Ibrahim) Ibrahim 1933

A. trialatus Kosanke 1950
Proprisporites laevigatus Neves 1961
Hymenospora multirugosa Peppers 1970
Laevigatosporites desmoinesensis (Wilson and Coe) Schopf, Wilson, and Bentall 1944
L. globosus Schemel 1951
L. medius Kosanke 1950
L. minimus (Wilson and Coe) Schopf, Wilson, and Bentall 1944
L. ovalis Kosanke 1950
L. punctatus Kosanke 1950
L. striatus Alpern 1959
L. vulgaris (Ibrahim) Ibrahim 1933
Renisporites confossus Winslow 1959
Latosporites minutus Bhardwaj 1957 (Bhardwaj, 1957a)
Punctatosporites granifer (Potonié and Kremp) Alpern and Doubinger 1973
P. minutus (Ibrahim) Alpern and Doubinger 1973
P. oculus Smith and Butterworth 1967
P. rotundus (Bhardwaj) Alpern and Doubinger 1973
Spinosporites exiguus Upshaw and Hedlund 1967
Tuberculatosporites robustus (Kosanke) Peppers 1970
Thymospora obscura (Kosanke) Wilson and Venkatachala 1963
T. pseudothiessenii (Kosanke) Wilson and Venkatachala 1963
T. thiessenii (Kosanke) Wilson and Venkatachala 1963
Dictyomonolites swadei Ravn 1986
Torispora securis (Balme) Alpern, Doubinger, and Horst 1965
Torispora verrucosa Alpern 1958
Vestispora clara (Venkatachala and Bharadwaj) Ravn 1986
V. costata (Balme) Spode (in Smith and Butterworth 1967)
V. fenestrata (Kosanke and Brokaw) Spode (in Smith and Butterworth, 1967)
V. foveata (Kosanke) Wilson and Venkatachala 1963
V. irregularis (Kosanke) Wilson and Venkatachala 1963
V. laevigata Wilson and Venkatachala 1963
V. magna (Butterworth and Williams) Spode (in Smith and Butterworth 1967)
V. pseudoreticulata Spode (in Smith and Butterworth 1967)
V. tortuosa (Balme) Spode (in Smith and Butterworth 1967)
V. wanlessii Peppers 1970
Florinites mediapudens (Loose) Potonié and Kremp 1956
F. millotti Butterworth and Williams 1954
F. similis Kosanke 1950
F. triletus Kosanke 1950
F. visendus (Ibrahim) Schopf, Wilson, and Bentall 1944
F. volans (Loose) Potonié and Kremp 1956
Wilsonites circularis (Guennel) Peppers and Ravn (in Ravn 1979)
W. delicatus (Kosanke) Kosanke 1959 (Kosanke, 1959b)
W. vesicatus (Kosanke) Kosanke 1959 (Kosanke, 1959b)
Potonieisporites elegans (Wilson and Kosanke) Wilson and Venkatachala 1964
P. solidus Ravn 1979
Pityosporites westphalensis Williams 1955
Vesicaspora wilsonii (Schemel) Wilson and Venkatachala 1963
Quasillinites diversiformis (Kosanke) Ravn and Fitzgerald 1982
Peppersites ellipticus Ravn 1979
Tinnulisporites cf. *microsaccus* Dempsey 1967
Trihyphaecites triangulatus Peppers 1970

APPENDIX B. LOCATIONS OF COALS DISCUSSED IN TEXT OR USED FOR TABLES

Maceration	Coal Name	Location	Maceration	Coal Name	Location
	ILLINOIS		2781	"Duck Creek" coal bed	Ctr. E. line,SE,SE,NE,Sec.30,T16N, R12E, Okmulgee County
573	Chapel (No. 8) Coal Member	SW,NE,SW,SW,Sec.7,T12N,R8W, Macoupin County			
577	Unnamed coal below Carlinville Limestone Member	SE,NE,NW,Sec26,T12N,R8W, Macoupin County	2797A	Dawson coal bed	Ctr. E. line,NW,Sec34,T27N,R15E, 3 mi. W of Delaware, Nowata County
			2797B	Tulsa coal bed	Ctr. E. line,NW,Sec34,T27N,R15E, 3 mi. W of Delaware, Nowata County
587	Gentry Coal Bed	Sec27,T11S,R10E, Hardin County			
693C	Lake Creek Coal Member	Old Ben Coal Company, Hole 49, NE,SW,SE,Sec18,T7S,R4E, Franklin County	2818	McAlester coal bed	NW,SW,Sec15,T1N,R8E, 6 mi. S of Ada, Pontotoc County
693D	Pond Creek Coal Member	Old Ben Coal Company, Hole 49, NE,SW,SE,Sec18,T7S,R4E, Franklin County	2827	Unnamed coal bed	NW,SE,NE,Sec23,T15N,R11E, 2 mi. NW Beggs, Okmulgee County
			2921	"Checkerboard" coal bed	NE,SW,NW,NE,Sec34,T19N,R21E, Tulsa County
693E	DeGraff Coal Member	Old Ben Coal Company, Hole 49, NE,SW,SE,Sec18,T7S,R4E, Franklin County	3035 B C	Coal in "Bostwick Member" Bottom half Upper half	NW,NW,SW,Sec15,T3S,R3E, Carter County
1178A	Unnamed coal bed	NW,NE,SE,Sec36,T17N,R4W, Montpelier quadrangle, Rock Island County	3121B	Hartshorne coal bed	N1/2,N1/2,Sec14,T14N,R19E, Harris County
2596 B C	Rock Branch Coal Member Lower 2 in. Upper 6 in.	SW,SW,NW,Sec16,T12N,R9W, Macoupin County	3153 A B	McAlester coal bed Lower 4 in. Upper 3 in.	W bank North Boggy Creek, NW of Koawa, T3N,R13E, Pittsburg County
2596D	Athensville Coal Member	SW,SW,NW,Sec16,T12N,R9W, Macoupin County		**MISSOURI**	
			772	Ovid coal bed	SE,NW,SW,Sec.30,T54N,R26W, Ray County
2611	Tunnel Hill Coal Bed	Abandoned mine,Nctr.NE,SWSW,Sec 17,T11S,R4E, Johnson County	778	Upper Locust Creek coal bed	SW,Sec27,T63N,R20W, Sullivan County
2824	Unnamed coal bed	SW,NW,NW,SE,Sec19,T11S,R4E, Johnson County	902	Drywood coal	NW,SW,Sec13,T13N,R31W, Barton County
2978	Tunnel Hill Coal Bed	SE,NW,SE,Sec23,T11S,R6E, Pope County	2779	Grain Valley coal	N. line,NW,Sec1,T44N,R32W, Case County
	INDIANA		2780	Laredo coal	SW,SW,Sec24,T60N,R23W, Grundy County
150	Pinnick Coal Member	Pinnick Quarry, SW,Sec32,T2N, R2W, Orange County			
151	French Lick Coal Member	Wortinger Whetstone Quarry, NW, Sec32,T2N,R2W, Orange County	2830	Grain Valley coal	S. line,SE,SE,NE,Sec35,T49N,R28W, Lafayette County
831	Unnamed coal below upper bench of West Franklin Limestone Member	Ctr.S1/2,Sec24,T7S,R12W, Posey County	2967 2968	Riverton coal, upper 9.5 in. Lower 1 ft, 2.5 in.	NW,NW,SW,Sec15,T33N,R29W, Barton County
2764	Pirtle Coal Member	SE,SE,SW,Sec15,T7N,R8W, Sullivan County	3118A 3118B	Riverton coal, lower 15 in. Upper 12 in.	NE,SE,Sec14,T31N,R31W, Barton County
	OKLAHOMA			**ARKANSAS**	
2714	Dawson coal bed	W bank Coal Creek, NW,SW, Sec 2,T17N,R12E, Tulsa County	1443	Baldwin coal	SW,NW,Sec12,T16N,R33W, Washington County
2715	Dawson coal bed	SW,SE,SE,Sec4,T24N,R15E, Rogers County	2793	Baldwin coal	S1/2,Sec26,T13N,R33W,W side Arkansas Highway 59, Washington County
				OHIO	
2725A	Coal below McAlester coal bed	Birch Land Company, Hole II, SW,SW,Sec32,T12N,R20E, Muskogee County	2643	Clarion coal bed	On State Route 93, just N of Olive Furnace, Oak Hill quadrangle, Lawrence County

APPENDIX B. LOCATIONS OF COALS DISCUSSED IN TEXT OR USED FOR TABLES (continued - page 2)

Maceration	Coal Name	Location
2788	Mahoning coal bed	Strip mine, Ctr.NW,SW,Sec12, Fox Township, Carroll County
2789	Mahoning coal bed	N1/2,SW,Sec29,W side Route 164, Washington Township, Columbiana County
2910 A B C	Upper Mercer coal bed Lower 10 in. Middle 10 in. Upper 7 in.	On State Route 60, 6 mi. N of Zanesville, Muskingum County
2911	Middle Mercer coal bed	On State Route 60, 6 mi. N of Zanesville, Muskingum County
2930	Unnamed coal	S bank Muskingum River, Putnam Hill, Zanesville, Muskingum County
2931	Brookville coal bed	S bank Muskingum River, Putnam Hill, Zanesville, Muskingum County
2964	Brush Creek coal bed	NW,NE,SW,Sec9,T2N,R3W, Cambridge quadrangle, Guernsey County
3052	Brush Creek coal bed	200 yds. S Route 682/56 intersection, Athens County
3100N	Mahoning coal bed	Along Route 11, Ctr. E line, SE,SE,Sec10,T6N,R2W, Columbiana County
3101C	Mason coal bed	7-11 strip mine, SE,SE,NW,Sec13, T10N,R2W, West Point quadrangle, Columbiana County

KANSAS

Maceration	Coal Name	Location
152	Riverton coal	NW,NE,Sec9, T32S,R25E, Cherokee County
2795	Unnamed coal below Sni Mills Limestone	Ctr., Sec5,T21S,R25E, at Trading Post, Union County
2796	Dawson coal	NE,NE,NE,Sec10,T33S,R18E, SW of Mound Valley, Labette County
2805	Unnamed coal above Worland Limestone	SE,SE,Sec34,T33S,R18E, 2 mi. NE Angola, Labette County
2808	"Hepler" coal	Ctr. W1/2,NE,Sec10,T33S,R18E, 1 mi. SW Mound Valley, Labette County
2809A	Dawson coal	Ctr. E line, SE,SE,SE,Sec19,T33S, R18E, 5 mi. SW Mound Valley, Labette County
2809B	"Hepler" coal	Ctr. E line, SE,SE,SE,Sec19,T33S, R18E, 5 mi. SW Mound Valley, Labette County
2831	"Hepler" coal	SE cor.,Sec19,T33S,R18E, Labette County
2892	"Hepler" coal	Ctr. N1/2NW,NE,Sec18,T32S,R19E, Labette County

TENNESSEE

Maceration	Coal Name	Location
981	Jordon (Peabody) coal bed	W of U.S. 25W in drainage of Lick Creek, Jellico East quadrangle, Campbell County
985	Jellico coal bed	Hatfield Creek overpass from Whisk Creek near Newcomb, Jellico West quadrangle, Campbell County
987	Unnamed coal bed	Whisk Branch, Jellico West quadrangle, Campbell County
988	Coal Creek Rider coal	Back of Careyville Railroad Station, Campbell County
989	Pewee Rider coal bed	Frozen Head Mountain near Petros, Briceville quadrangle, Morgan County, elevation 2,580 ft.
995	Unnamed coal bed	Frozen Head Mountain near Petros, Briceville quadrangle, Morgan County, 20 ft below Frozen Head coal, elevation 2,390 ft.
996 A B	Pewee coal bed Upper 18 in. Lower 21 in.	Frozen Head Mountain near Petros, Briceville quadrangle, Morgan County, 150 ft below fossiliferous olive shale, elevation 2,560 ft.
997 A B	Frozen Head coal bed Upper 8.5 in. Lower 24.5 in	Frozen Head Mountain near Petros, Briceville quadrangle, Morgan County, elevation 2,410 ft.
998 A B	Unnamed coal bed Upper 12 in. Bottom 6 in.	Frozen Head Mountain near Petros, Briceville quadrangle, Morgan County, 35 ft above Pewee coal
1017 A B C	Joyner coal bed Upper 8 in. Middle 8 in Bottom 13 in.	Frozen Head Mountain near Petros, Briceville quadrangle, Morgan County, just above sandstone near base of Indian Bluff Group, elevation 1,700 ft.
2665 A B C D	Big Mary coal bed Upper 10 in. Upper middle 2 in. Lower middle 12 in. Lower 12 in.	Frozen Head Mountain near Petros, Briceville quadrangle, Morgan County, near base of Redoak Mountain Group, elevation 2,270 ft.
990	Blue Gem coal bed	Loudan Coal Mine, Burnt Pone Creek, 1.75 mi NNW Newcomb, Jellico West quadrangle, Campbell County
1003	Poplar Creek coal bed	On State Highway 62, W side Corbin Hill, 2 mi. E of Wartburg, Morgan County.
1004	Hooper coal bed	Brickyards at Oliver Springs, Briceville quadrangle, junction of Morgan, Anderson, and Roane Counties
1005	Ant coal bed	Railroad cut between Coalfield and Stevens, Morgan County

APPENDIX B. LOCATIONS OF COALS DISCUSSED IN TEXT OR USED FOR TABLES (continued - page 3)

Maceration	Coal Name	Location	Maceration	Coal Name	Location
1006	Windrock coal bed	Braden Coal Mine, Graves Gap, Briceville quadrangle, Anderson County	1961D	Unnamed coal bed	135 ft, 4 in—135 ft, 5.5 in.
A	Upper 12 in.		1961E	Unnamed coal bed	134 ft, 10.5 in—135 ft, 0.5 in.
B	Upper middle 12 in.				
C	Middle 12 in.		1961F	Unnamed coal bed	130 ft, 11 in.—131 ft, 1 in.
D	Lower middle 12 in.				
E	Lower 8 in.		1961G	Unnamed coal bed	112 ft, 2 in.—112 ft, 10 in.
1009	Clifty coal bed	Near Clifty, Crossville quadrangle, White County projection into western Cumberland County	1987	Princess No. 5 coal bed	NW corner intersection Highway 60 and 64, 5 mi. SW Ashland, Boyd County.
A	Upper 12 in.		A	Lower 16 in.	
B	Upper middle 12 in.		B	Lower middle 8 in.	
C	Lower middle 12 in.		C	Middle 5 in.	
D	Lower 12 in.		D	Upper middle 3 in.	
			E	Upper 5 in.	
1012	Morgan Springs coal bed	Along Highway 30, just east of Morgan Springs, Sequatchie Valley quadrangle, Rhea County	2110A	Unnamed coal bed	435 ft, 10 in.—436 ft, 6in.; Hole C, Midcontinental Coal and Transport Company, NW,NE,SW,NE, 20-M-18, Crittendon County
A	Upper 12 In.				
B	Lower 12 in.				
1015	Upper Pioneer coal bed	Just W of Pioneer, Briceville quadrangle, Campbell County	2110B	Unnamed coal bed	390 ft, 4 in.—390 ft, 8 in.
1011	Coal Creek coal bed	Above Coal Creek near Lake City, base of Cross Mountain, Briceville quadrangle, Anderson County	2111B	Unnamed coal bed	44 ft, 9in.—45 ft, 9 in.; Hole F, Midcontinent Coal and Transport Company, NE,NE,NW,NW, 11-M-18, Crittendon County.
A	Upper 10 in.				
B	Upper middle 12 in.				
C	Middle 12 in.				
D	Lower middle 12 in.		2111C	Unnamed coal bed	291 ft, 2 in.—291 ft, 6 in.
E	Lower 4 in.				
1025	Unnamed coal bed	Highest coal on Cross Mountain, 3.5 mi. SW Lake City, Briceville quadrangle, Anderson County	2111D	Unnamed coal bed	255 ft, 2.5 in.—255 ft, 11 in.
A	Upper 9 in.				
B	Lower 9 in.		2111E	Unnamed coal bed	235 ft, 3 in.—235 ft, 9 in.
1026	Unnamed coal bed	Cross Mountain, 3.5 mi. SW Lake City, Briceville quadrangle, Anderson County, 137 ft below top bench.	2113	Smith coal bed	Mine No. 2, Midcontinental Coal and Transport Company, SE,NW,SE, 20-M-18, Crittendon County.
			A	Bottom 8 in.	
			B	Lower middle 8 in.	
1027	Wildcat coal bed? (Unnamed coal bed)	Cross Mountain, 3.5 mi. SW Lake City, Briceville quadrangle, Anderson County. 165 ft. below top bench.	C	Upper middle 8 in.	
			C	Upper 8 in.	
1028	Cold Gap coal bed	Cross Mountain, 3.5 mi. SW Lake City, Briceville quadrangle, Anderson County, 187 ft. below top bench.	2168C	Unnamed coal bed	1,100 ft W line, 3,400 ft N line, 19-I-34, 1 mi. W Cool Spring Church, Green River Parkway, Butler County
1029	Rock Springs coal bed	Cross Mountain, 3.5 mi. SW Lake City, Briceville quadrangle, Anderson County, 437 ft. below top bench.	2198	Tunnel coal bed	16,600 ft W line, 1,800 ft N line, Middlesboro North quadrangle, Bell County
A	Upper 4 in.				
B	Lower 5 in.				
1030	Pine Bald coal bed	Cross Mountain, 3.5 mi. SW Lake City, Briceville quadrangle, Anderson County, elevation, 2,780 ft.	2255	Bell coal bed	Strip mine 0.5 mi. SE Bell Mines Church, 21-M-18, 400 ft S line, 1,000 ft W line, Crittendon County
A	Upper 12 in.				
B	Lower 12 in.				
			2461	"Battery Rock" coal bed	500 ft S line, 2,250 ft E line, 24-M-18, S of mouth of Tradewater River, Union County
		KENTUCKY	2567F	Unnamed coal equivalent Rock Island (No. 1) Coal	392.7-394.3 ft, Kentucky Geological Survey, Hole Gil 15, 2,100 ft W line, 1,950 ft S line, Dekoven quadrangle, Union County
991	River Gem coal bed	Saxton Mine, 3 mi. N of Jellico, Just N of Kentucky-Tennessee line, Whitley County.			
A	Upper 16 in.				
B	Lower 14.5 in.		2666	Hazard No. 4 coal bed	Messer Branch, 1 mi SW Hazard, 2-I-76, Hazard South quadrangle, Perry County
1961B	Unnamed coal bed	146 ft, 0 in—146 ft, 9.5 in., Hole H Midcontinental Coal and Transport Company, SE,SW,SW,SE,20-M-18, Crittendon County.	E	Lower 9 in.	
			F	Middle 22 in.	
			G	Upper 14.5 in.	
1961C	Unnamed coal bed	136 ft, 2.5 in.—136 ft, 4 in.	2666H	Hazard No. 4 rider coal bed	Messer Branch, 1 mi SW Hazard, 2-I-76, Hazard South quadrangle, Perry County

APPENDIX B. LOCATIONS OF COALS DISCUSSED IN TEXT OR USED FOR TABLES (continued - page 4)

Maceration	Coal Name	Location	Maceration	Coal Name	Location
2666 J K L M	Hazard No. 6 coal bed Lower 5 in. Lower middle 15.5 in. Upper middle 8.5 in. Upper 7.5 in.	Messer Branch, 1 mi SW Hazard, 2-I-76, Hazard South quadrangle, Perry County	2712	Upper Elkhorn coal bed	37°40'15"N, 82°44'32"W, at junction of Levisa Fork and Brandykeg Creek at Lancer, Lancer quadrangle, Floyd County
2666 N O P	Hazard No. 7 coal bed Lower 12.5 in. Middle 15.5 in. Upper 18 in.	Messer Branch, 1 mi SW Hazard, 2-I-76, Hazard South quadrangle, Perry County	2737 A B C	Princess No. 4 coal bed Lower 15 in. Middle 14 in. Upper 9 in.	4,900 ft. S line, 2,900 ft. E line, 25-X-80, Oldtown quadrangle, Greenup County
2671 A B C D	Hazard No. 8 coal bed Lower 4.75 in. Lower middle 3.5 in. Upper middle 1.4 in. Upper 7 in.	Along Buffalo Creek on road to Four Seams Coal Company, Perry County	2738	Gray Hawk coal bed	3,500 ft. S line, 3,350 ft. E line, 23-M-69, Heidelberg quadrangle, Lee County
2669	Hamlin coal bed	E. side Highway 15, SE corner Hayden quadrangle, Perry County	2741	Beattyville coal bed	2,400 ft. S line, 1,900 ft. E line, 3-M-70, Beattyville quadrangle, Lee County
2675A	Little Fire Clay coal bed	Greater Branch along U.S. 460, Morgan County	2742	Barren Fork coal bed	3,950 ft. S line, 2,600 ft. E line, 22-E-61, Webers quadrangle, McCreary County
2675B	Copeland coal bed	Greater Branch along U.S. 460, Morgan County	2755	Hagy coal bed	3,800 ft. S line, 2,750 ft. E line, 7-J-87, Elkhorn City quadrangle, Pike County
2678 A B C D	Alma coal bed Upper 28 in. Upper middle 9 in. Lower middle 17 in. Lower 16 in.	W. side Tug Fork, directly across from Borderland, West Virginia, Pike County	2760	Amburgy coal bed	0.5 mi. E of Hayden, Leslie County
			2776	Stearns coal bed	6,000 ft. S line, 800 ft E line, Nevelsville quadrangle, Wayne County
2699	Pond Creek Rider Coal bed	3,300 ft. S line, 4,550 ft. E line, 2-L-86, Lick Creek quadrangle, Pike County	2783	Mason coal bed	2,750 ft. S line, 2,300 ft E line, 25-C-72, Varilla quadrangle, Bell County
2700	Pond Creek coal bed	5,000 ft. S line, 500 ft. E line, 13-P-86, Nangatuck quadrangle, Martin County		**PENNSYLVANIA**	
			2816	Upper Freeport coal bed	Creighton Mine at Creighton, Pittsburg Plate Glass Company, Allegheny County
2710	Princess No. 6 coal bed	3,950 ft. S line, 1,000 ft. E line, 1-W-82, Ashland quadrangle, Boyd County		**WEST VIRGINIA**	
2711	Haddix coal bed	2,700 ft. S line, 4,600 ft. E line, 11-G-71, Creekville quadrangle, Clay County	2763	Lower War Eagle coal bed	5 mi. NW of Gilbert on Route 52, Gilbert quadrangle, Mingo County

REFERENCES CITED

Adams G. I., 1904, Zinc and lead deposits of northern Arkansas: U.S. Geological Survey Professional Paper 24, 118 p.

Aizenverg, D. E., and 10 others, 1978, Carboniferous sequence of the Donets Basin as the standard section of the Carboniferous, *in* Meyen, S. V., and others, eds., General problems of the Carboniferous stratigraphy: Compte Rendu, Huitième Congrès International de Stratigraphie et de Géologie du Carbonifère, Moscow, 1975, v. 1, p. 158–168.

Alpern, B., 1958, Description de quelques microspores du Permo-Carbonifère français: Revue de Micropaléontologie, v. 1, p. 75–86.

Alpern, B., 1959, Contribution à l'étude palynologique et pétrographique des charbons français [Ph.D. thesis]: Paris, L'Université de Paris, 314 p.

Alpern, B., 1960, Quelques problèmes actuels de la palynologie houillère: Compte Rendu, Quatrième Congrès l'advancement des Études de Stratigraphie et de Géologie du Carbonifère, Heerlen, 1958, p. 13–24.

Alpern, B., 1963, Coupe palynologique du Westphalien du bassin houiller de Lorraine: Paris, Académie des Sciences, Comptes Rendus, v. 256, p. 5170–5172.

Alpern, B., and Doubinger, J., 1973, Les microspores monolètes de Paleozoique: Commission Internationale de Microflore du Paléozoique, fascicule 6, p. 1–103.

Alpern, B., and Liabeuf, J. J., 1966, Zonation palynologique du bassin houiller lorrain: Zeitschrift Deutschen Geologischen Gesellschaft, Band 117, p. 162–177.

Alpern, B., and Liabeuf, J. J., 1967, Considerations palynologiques sur le Westphalien et le Stephanien: propositions pour une parastratotype: Paris, Académie des Sciences, Comptes Rendus, v. 265, p. 840–843.

Alpern, B., and Liabeuf, J. J., 1969, Palynological considerations on the Westphalian and Stephanian: proposition for a parastratotype: Compte Rendu, Sixième Congrès International de Stratigraphie et de Géologie du Carbonifère, Sheffield, 1967, v. 1, p. 109–114.

Alpern, B., Girardeau, J., and Trolard, F., 1960, Répartition stratigraphique de quelques microspores du Carbonifère supérieur France: International Commission on Coal Petrology, Proceedings, No. 3, p. 173–176.

Alpern, B., Balme, B., Doubinger, J., Goubin, N., Grebe, H., Navale, G., and Pierart, P., 1964, La stratigraphie palynologique du Stephanien et du Permien: Compte Rendu, Cinquième Congrès International de Stratigraphie et de Géologie du Carbonifère, Paris, 1963, v. 3, p. 1119–1129.

Alpern, B., Doubinger, J., and Horst, U., 1965, Révision du genre *Torispora* Balme: Pollen et Spores, v. 7, p. 565–572.

Alpern, B., Lachkar, G., and Liabeuf, J. J., 1967, Le bassin houiller lorrain peut-il fournir un stratotype pour le Westphalien supérieur?: Review of Palaeobotany and Palynology, v. 5, p. 75–91.

Andresen, M. J., 1956, Subsurface stratigraphy and sedimentation of the West Franklin–Cutler limestone zones of southern Illinois [M.S. thesis]: Urbana, University of Illinois, 31 p.

Arkle, T., Jr., 1969, The configuration of the Pennsylvanian and Dunkard (Permian?) strata in West Virginia: A challenge to classical concepts, *in* Some Appalachian coals and carbonates, models of ancient shallow-water deposition (Geological Society of America, Coal Division, Preconvention Field Trip 1969): Morgantown, West Virginia, West Virginia Geological and Economic Survey, p. 55–88.

Arkle, T., Jr., and 9 others, 1979, The Mississippian and Pennsylvanian (Carboniferous) Systems in the United States—West Virginia and Maryland: U.S. Geological Survey Professional Paper 1110-D, 35 p.

Arndt, H. H., 1979, Middle Pennsylvanian Series in the proposed Pennsylvanian System stratotype, *in* Englund, K. J., Arndt, H. H., and Henry, T. W., eds., Proposed Pennsylvanian System stratotype Virginia and West Virginia (Ninth International Congress of Carboniferous Stratigraphy and Geology, Guidebook, Field Trip No. 1): American Geological Institute Guidebook Series 1, p. 73–80.

Artüz, S., 1957, Die *Sporae dispersae* der Türkischen Steinkohle von Zonguldak-Gebiet: Revue de la Faculté des Sciences de l'Université d'Istanbul, serie B, tome 22, fascicule 4, p. 239–263.

Atherton, E., and Palmer, J. E., 1979, The Mississippian and Pennsylvanian (Carboniferous) Systems in the United States—Illinois: U.S. Geological Survey Professional Paper 1110-L, 42 p.

Balme, B. E., and Hassell, C. W., 1962, Upper Devonian spores from the Canning Basin, Western Australia: Micropaleontology, v. 8, p. 1–28.

Bambach, R. K., Scotese, C. R., and Ziegler, A. M., 1980, Before Pangea: The geographies of the Paleozoic world: American Scientist, v. 68, p. 26–38.

Barrick, J. E., and Boardman, D. R., II, 1989, Stratigraphic distribution of morphotypes of *Idiognathodus* and *Streptognathodus* in Missourian–lower Virgilian strata, north-central Texas, *in* Boardman D. R., II, Barrick, J. E., Cocke, J., and Nestell, M. K., eds., Middle and Late Pennsylvanian chronostratigraphic boundaries in North-Central Texas: glacial-eustatic events, biostratigraphy and paleoecology: Texas Tech University Studies in Geology, v. 2, p. 167–189.

Berner, R. A., 1991, A model for atmospheric CO_2 over Phanerozoic time: American Journal of Science, v. 291, p. 339–376.

Bhardwaj, D. C., 1957a, The palynological investigations of the Saar coals: Palaeontographica B, v. 101, p. 73–125.

Bhardwaj, D. C., 1957b, The spore flora of the Velener Schichten (Lower Westphalian D) in the Ruhr Coal Measures: Palaeontographica B, v. 102, p. 110–138.

Bharadwaj, D. C., 1960, Sporological evidence on the boundaries of the stratigraphical subdivisions in the Upper Pennsylvanian strata of Europe and North America: Compte Rendu, Quatrième Congrès Advancement Etudes Stratigraphic Géologie Carbonifère, Heerlen, 1958, p. 33–39.

Bless, M. J. M., Loboziak, S., and Streel, M., 1977, An upper Westphalian C "hinterland" microflora from the Haaksbergen-1 borehole (Netherlands): Mededelingen van's Rijks Geologische Dienst, new series 28, p. 135–147.

Boardman, D. R., II, and Heckel, P. H., 1989, Glacial-eustatic sea-level curve for early Late Pennsylvanian sequence in north-central Texas and bio-stratigraphic correlation with curve for midcontinent North America: Geology, v. 17, p. 802–805.

Boardman, D. R., II, and Mapes, R. H., 1984, Preliminary placement of the Desmoinesian-Missourian boundary utilizing ammonoid cephalopods, *in* Bennison, A. P., ed., Pennsylvanian source beds of northeastern Oklahoma and adjacent Kansas: Tulsa Geological Society Field Trip Guidebook, p. 54–58.

Boardman, D. R., II, Barrick, J. E., and Heckel, P. H., 1989a, Proposed redefinition of the Missourian-Virgilian Stage boundary (Late Pennsylvanian), Midcontinent North America: Geological Society of America Annual meeting Abstracts with Programs, v. 21, no. 6, p. 168.

Boardman, D. R., II, Barrick, J. E., Heckel, P. H., and Nestell, M. K., 1989b, Upper Pennsylvanian chronostratigraphic subdivisions of the North American Midcontinent, *in* Boardman, D. R., II, Barrick, J. E., Cocke, J., and Nestell, M. K., eds., Middle and Late Pennsylvanian chronostratigraphic boundaries in north-central Texas: Glacial-eustatic events, biostratigraphy, and paleoecology: Texas Technical University Studies in Geology, v. 2, p. 1–16.

Boardman, D. R., II, Mapes, R. H., and Work, D. M., 1989c, Early Missourian ammonoids from the North American Midcontinent, *in* Boardman, D. R., II, Barrick, J. E., Cocke, J., and Nestell, M. K., eds., Middle and Late Pennsylvanian chronostratigraphic boundaries in north-central Texas: Glacial-eustatic events, biostratigraphy and paleoecology: Texas Technical University Studies in Geology, v. 2, p. 151–166.

Boardman, D. R., II, Heckel, P. H., Barrick, J. E., Nestell, M. K., and Peppers, R. A., 1990, Middle-Upper Pennsylvanian chronostratigraphic boundary in the Midcontinent region of North America, *in* Brenckle, P. L., and Manger, W. L., eds., Intercontinental correlation and division of the Carboniferous System: Courier Forschungsinstitut Senckenberg, v. 130, p. 319–337.

Bode, H. H., 1958, Die floristische Gliederung des Oberkarbons der Vereinigten Staaten von Nordamerika: Zeitschrift der Deutschen Geologischen Gesellschaft, v. 110, p. 217–259.

Bode, H. H., 1975, The stratigraphic position of the Dunkard, *in* Barlow, J. A., ed., The age of the Dunkard: Proceedings of the First I. C. White Memorial Symposium, Morgantown, West Virginia, West Virginia Geological Economic Survey, p. 143–154.

Bordeau, K. V., 1964, Palynology of the Drywood coal (Pennsylvanian) of Oklahoma [M.S. thesis]: Norman, University of Oklahoma, 207 p.

Bouroz, A., 1978, Presidential address, IUGS Subcommission on Carboniferous Stratigraphy, *in* Meyen, S. V., and others, eds., General problems of the Carboniferous stratigraphy: Compte Rendu, Huitième Congrès International de Stratigraphie et de Géologie du Carbonifère, Moscow, 1975, v. 1, p. 22–26.

Bouroz, A., Gras, H., and Wagner, R. H., 1970, A propos de la limite Westphalien-Stéphanien et du Stéphanien inférieur: Colloque sur la Stratigraphie du Carbonifère, Université de Liège, v. 55, p. 205–220.

Bouroz, A., Einor, O. L., Gordon, M., Meyen, S. V., and Wagner, R. H., 1978, Proposals for an international chronostratigraphic classification of the Carboniferous, *in* Meyen, S. V., and others, eds., General problems of the Carboniferous stratigraphy: Compte Rendu, Huitième Congrès International de Stratigraphie et de Géologie du Carbonifère, Moscow, 1975, v. 1, p. 36–69.

Bradley, W. H., 1956, Use of series subdivisions of the Mississippian and Pennsylvanian System: American Association of Petroleum Geology Bulletin 40, p. 2284–2285.

Bristol, H. M., and Howard, R. H., 1971, Paleogeologic map of the sub-Pennsylvanian Chesterian (Upper Mississippian) surface in the Illinois Basin: Illinois State Geological Survey Circular 458, 14 p.

Budyko, M. I., 1982, The earth's climate: past and future (International Geophysics Series 29): New York, Academic Press, 309 p.

Butterworth, M. A., 1964, Miospore distribution in the Namurian and Westphalian: Compte Rendu, Cinquième Congrès International de Stratigraphie et de Géologie du Carbonifère, Paris 1963, v. 3, p. 1115–1118.

Butterworth, M. A., 1969, Microfloras of the Upper Carboniferous: Compte Rendu, Cinquième Congrès International de Stratigraphie et de Géologie du Carbonifère, Sheffield, 1967, v. 1, p. 59–70.

Butterworth, M. A., and Millott, J. O., 1954, Microspore distribution in the seams of the North Staffordshire, Cannock Chase and North Wales coalfields: London, Institute of Mining Engineers, Transactions, v. 114, p. 501–520.

Butterworth, M. A., and Smith, A. H. V., 1976, The age of the British Upper Coal Measures with reference to their miospore content: Review of Palaeobotany and Palynology, v. 22, p. 281–306.

Butterworth, M. A., and Williams, R. W., 1958, The small spore floras of the coals in the Limestone Coal Group and Upper Limestone Group of the Lower Carboniferous of Scotland: Royal Society of Edinburgh, Transactions, v. 63, p. 353–392.

Butterworth, M. A., Jansonius, J., Smith, A. H. V., and Staplin, F. L., 1964, *Densosporites* (Berry) Potonié and Kremp, and related genera. Report of the Commission Internationale de Microflore du Paleozoic Working Group No. 2: Compte Rendu, Cinquième Congrès International de Stratigraphie et de Géologie du Carbonifère, Paris, 1963, v. 3, p. 1049–1057.

Butterworth, M. A., Mahdi, S. A., and Nader, A. D., 1988, Miospores from the Westphalian A and B of northern England: Pollen et Spores, v. 30, p. 57–80.

Cady, G. H., 1933, The physical constitution of Illinois coal and its significance in regard to utilization: Illinois Mining Institute, Proceedings, p. 95–111.

Campbell, M. R., and Mendenhall, W. C., 1896, Geologic section along the New and Kanawha rivers in West Virginia: U.S. Geological Survey Annual Report 17, p. 473–511.

Cecil, C. B., 1990, Paleoclimate controls on stratigraphic repetition of chemical and siliciclastic rocks: Geology, v. 18, p. 533–536.

Cecil, C. B., Stanton, R. W., Neuzil, S. G., Dulong, F. T., Ruppert, L. F., and Pierce, B. S., 1985, Paleoclimate controls on late Paleozoic sedimentation and peat formation in the central Appalachian basin (U.S.A.): International Journal of Coal Geology, v. 5, p. 195–230.

Châteauneuf, J. J., 1973, Palynologie des faisceaux productifs du bassin central des Asturies (Espagne): Compte Rendu, Septième Congrès International de Stratigraphie et de Géologie du Carbonifère, Krefeld, 1971, v. 2, p. 297–321.

Cheney, M. G., 1940, Geology of north-central Texas: American Association of Petroleum Geologists Bulletin 24, p. 65–118.

Cheney, M.G., and six others, 1945, Classification of Mississippian and Pennsylvanian rocks of North America: American Association of Petroleum Geologists Bulletin, v. 29, p. 125–169.

Chesnut, D. R., Jr., and Cobb, J. C., 1989, Cycles in the Pennsylvanian rocks of the Central Appalachian Basin: Geological Society of America Abstracts with Programs, v. 21, no. 6, p. A52.

Christopher, C. C., Hoare, R. D., and Sturgeon, M. T., 1990, Pennsylvanian hollinacean and kirkbyacean ostracodes from the Appalachian Basin: Journal of Paleontology, v. 64, p. 967–987.

Clayton, G., Coquel, R., Doubinger, J., Gueinn, K. J., Loboziak, S., Owens, B., and Streel, M., 1977, Carboniferous miospores of western Europe: Report of Commission Internationale de Microflore du Paléozoic Working Group on Carboniferous Stratigraphic Palynology: Mededelingen van's Rijks Geologische Dienst, v. 29, 11 p.

Cleal, C. J., 1984, The Westphalian D floral biostratigraphy of Saarland (Federal Republic of Germany) and a comparison with that of South Wales: Geological Journal, v. 19, p. 327–351.

Clendening, J. A., 1974, Palynological evidence for a Pennsylvanian age assignment of the Dunkard Group in the Appalachian Basin: Part II: West Virginia Geological and Economic Survey, Coal Geology Bulletin 3, 105 p.

Clendening, J. A., and Gillespie, W. H., 1964, Characteristic small spores of the Pittsburgh coal in West Virginia and Pennsylvania: West Virginia Academy of Science Proceedings, v. 35, p. 141–150.

Cline, L. M., 1941, Traverse of upper Des Moines and lower Missouri Series from Jackson County, Missouri to Appanoose County, Iowa: American Association of Petroleum Geologists Bulletin, v. 25, p. 23–72.

Cline, L. M. and Greene, F. C., 1950, A stratigraphic study of the Upper Marmaton and lowermost Pleasanton groups, Pennsylvanian of Missouri: Missouri Geological Survey and Water Resources Report of Investigations 12, 74 p.

Clopine, W. W., 1986, The lithostratigraphy, biostratigraphy, and depositional history of the Atokan Series (Middle Pennsylvanian) in the Ardmore Basin, Oklahoma [M.S. thesis]: Norman, University of Oklahoma, 161 p.

Clopine, W. W., 1991, Lithostratigraphic and biostratigraphic analysis of the Atokan Series (Middle Pennsylvanian) in the Ardmore Basin, Oklahoma: Compass, v. 68, p. 221–232.

Cobb, J. C., Chesnut, D. R., Hester, N. C., Hower, J. C., Rice, C. L., and Jennings, J. R., 1981, Coal and coal-bearing rocks of eastern Kentucky: Annual Geological Society of America Coal Division Field Trip Guidebook, 169 p.

Collins, H. R., 1979, The Mississippian and Pennsylvanian (Carboniferous) Systems in the United States—Ohio: U.S. Geological Survey Professional Paper 110-E, 26 p.

Connolly, W. M., and Stanton, R. J., Jr., 1992, Interbasinal cyclostratigraphic correlation of Milenkovitch band transgressive-regressive cycles: Correlation of Desmoinesian-Missourian strata between southeastern Arizona and the midcontinent of North America: Geology, v. 20, p. 999–1002.

Cooper, C. L., 1946, Pennsylvanian ostracodes of Illinois: Illinois State Geological Survey Bulletin 70, 177 p.

Coquel, R., 1976, Etude palynologique de la Série houillère dans l'unité de production de Valenciennes du Basin le houillère du Nord de la France: Palaeontographica B, v. 156, p. 12–64.

Coquel, R., Loboziak, S., and Lemoigne, Y., 1970, Confirmation de l'âge Westphalien du Houiller de Le Plessis (Manche) d'après l'étude de quelques échantillions de charbons: Annales de Société Géologique du Nord, v. 90, p. 15–21.

Coquel, R., Doubinger, J., and Loboziak, S., 1976, Les microspores-guides du Westphalien à l'Autuien d'Europe occidentále: Revue de Micropaléontology, v. 18, p. 200–212.

Coquel, R., Loboziak, S., Owens, B., and Teteriuk, V. K., 1984, Comparaison entre la distribution des principales microspores—guide du Namurian et du Westphalien en Europe occidentale et dans le bassin du Donetz (URSS), *in* Sutherland, P. K., and Manger, W. L., eds., Biostratigraphy: Compte Rendu, Neuvième Congrès International de Stratigraphie et de Géologie du Carbonifère, Washington and Champaign-Urbana, v. 2, p. 443–446.

Corsin, P., Corsin, P., and Guerrier, R., 1968, A propos de la limite Westphalien-Stephanien: Paris, Académie des Sciences, Comptes Rendus, ser. D, v. 266, p. 1373–1378.

Cropp, F. W., III, 1958, Pennsylvanian spore succession in Tennessee [Ph.D. thesis]: Champaign, University of Illinois, 75 p.

Cropp, F. W., III, 1963, Pennsylvanian spore succession in Tennessee: Journal of Paleontology, v. 37, p. 900–916.

Cross, A. T., 1947, Spore floras of the Pennsylvanian of West Virginia and Kentucky: Journal of Geology, v. 55, p. 285–308.

Cross, A. T., 1952, The geology of the Pittsburgh coal: stratigraphy, petrology, origin and composition, and geologic interpretations of mining problems: Proceedings, Conference on Origin and Constitution of Coal, 2nd, Crystal Cliffs, Nova Scotia, p. 32–99.

Cross, A. T., 1992, Palynology of the earliest Pennsylvanian rocks, southern Indiana: American Association of Stratigraphic Palynologists, Abstracts, v. 16, p. 215–216.

Cross, A. T., and Schemel, M. P., 1952, Representative microfossil floras of some Appalachian coals: Compte Rendu, Troisième Congrès International de Stratigraphie et de Géologie du Carbonifère, Heerlen, 1951, v. 1, p. 123–130.

Darrah, W. C., 1970, A critical review of the Upper Pennsylvanian floras of the eastern United States with notes on the Mazon Creek flora of Illinois: Published by the author, 220 p.

Davis, P. N., 1961, Palynology of the Rowe coal (Pennsylvanian) of Oklahoma [M.S. thesis]: Norman, University of Oklahoma, 153 p.

Dawson, C. A., 1989, Notes on the palynology of the Morrowan reference section, Evansville Mountain, Arkansas, *in* Miller, M. A., Eames, L. E., and Prezbindowski, D. R., eds., Upper Mississippian and Lower Pennsylvanian lithofacies and palynology from northeastern Oklahoma: A field excursion: American Association of Stratigraphic Palynologists Field Trip Guidebook, p. 62–64.

Dawson, J. W., 1854, On the coal measures of South Joggins, Nova Scotia: Geological Society of London Quarterly Journal, v. 10, p. 1–42.

Dempsey, J. E., 1964, A palynological investigation of the Lower and Upper McAlester coals (Pennsylvanian) of Oklahoma [Ph.D. thesis]: Norman, University of Oklahoma, 124 p.

Dempsey, J. E., 1967, Sporomorphs from Lower and Upper McAlester coals (Pennsylvanian) of Oklahoma: An interim report: Review of Palaeobotany and Palynology, v. 5, p. 111–118.

Diez, M. D. C. R., and Cramer, F. H., 1979, Illustration of miospores from the Westphalian-Stephanian transition (late Carboniferous) of Asturias, Spain: Palinologia, v. 1, p. 179–209.

DiMichele, W. A., 1985, *Diaphorodendron*, gen nov., a segregate from *Lepidodendron* (Pennsylvanian age): Systematic Botany, v. 10, p. 453–458.

DiMichele, W. A., and Hook, R. W., 1992, Paleozoic terrestrial ecosystem, *in* Behrensmeyer, A. K., Damuth, J. D., DiMichele, W. A., Potts, R., Sues, H., and Wing, S. L., eds., Terrestrial ecosystems through time, evolutionary paleoecology of terrestrial plants and animals: Chicago and London, The University of Chicago Press, p. 205–325.

DiMichele, W. A., and Phillips, T. L., 1994, Paleobotanical and paleoecological constraints on models of peat formation in the Late Carboniferous of Euramerica: Palaeogeography, Palaeoclimatology, Palaeoecology, v. 106, p. 39–90.

DiMichele, W. A., Mahaffy, J. F., and Phillips, T. L., 1979, Lycopods of Pennsylvanian age coals: *Polysporia*: Canadian Journal of Botany, v. 57,

p. 1740–1753.

DiMichele, W. A., Phillips, T. L., and Peppers, R. A., 1985, The influence of climate and depositional environment in the distribution and evolution of Pennsylvanian coal-swamp plants, *in* Tiffney, B. H., ed., Geological factors and the evolution of plants: New Haven, Connecticut, Yale University Press, p. 223–256.

DiMichele, W. A., Phillips, T. L., and Olmstead, R. G., 1987, Opportunistic evolution: Abiotic environmental stress and the fossil record of plants: Review of Palaeobotany and Palynology, v. 50, p. 151–178.

Dix, E., 1934, The sequences of floras in the Upper Carboniferous with special reference to South Wales: Royal Society of Edinburgh Transactions, v. 57, p. 789–838.

Donaldson, A. C., Renton, J. J., and Presley, M. W., 1985, Pennsylvanian deposystems and paleoclimates of the Appalachians: International Journal of Coal Geology, v. 5, p. 167–193.

Douglass, R. C., 1979, The distribution of fusulinids and their correlation between the Illinois Basin and the Appalachian Basin, *in* Palmer, J. E., and Dutcher, R. R., eds., Depositional and structural history of the Pennsylvanian System of the Illinois Basin: Part 2: Invited papers: Ninth Congress of Carboniferous Stratigraphy and Geology, Champaign-Urbana, Field Trip 9, part 2; Illinois State Geological Survey Guidebook 15a, p. 15–20.

Douglass, R. C., 1987, Fusulinid biostratigraphy and correlations between the Appalachian and Eastern Interior basins: U.S. Geological Survey Professional Paper 1451, 95 p.

Douglass, R. C., and Nestell, M. K., 1984, Fusulinids of the Atoka Formation, Lower-Middle Pennsylvanian, south-central Oklahoma, *in* Sutherland, P. K., and Manger, W. L., eds., The Atokan Series (Pennsylvanian) and its boundaries—A symposium: Oklahoma Geological Survey Bulletin, 136, p. 19–39.

Droste, J. B., and Keller, S. J., 1989, Development of the Mississippian-Pennsylvanian unconformity in Indiana: Indiana Geological Survey Occasional Paper 55, 11 p.

Dunbar, C. O., and Condra, G. E., 1932, Brachiopods of the Pennsylvanian System of Nebraska: Nebraska Geological Survey, Series 2, Bulletin 5, 337 p.

Dunbar, C. O., and Henbest, L. G., 1942, Pennsylvanian Fusulinidae of Illinois: Illinois State Geological Survey Bulletin 67, 218 p.

Dunn, D. L., 1976, Biostratigraphic problems of Morrowan and Derryan (Atokan) strata in the Pennsylvanian System of western United States: Geological Society of America Bulletin 87, p. 641–645.

Durden, C. J., 1969, Pennsylvanian correlations using blattoid insects: Canadian Journal of Earth Sciences, v. 6, p. 1159–1177.

Durden, C. J., 1984a, North American provincial insect ages for the continental last half of the Permian, *in* Sutherland, P. K., and Manger, W. L., eds., Biostratigraphy: Compte Rendu, Neuvième Congrès International de Stratigraphie et de Géologie du Carbonifère, Washington and Champaign-Urbana, 1979, v. 2, p. 606–612.

Durden, C. J., 1984b, Age zonation of the Early Pennsylvanian using fossil insects, *in* Sutherland, P. K., and Manger, W. L., eds., The Atokan Series (Pennsylvanian) and its boundaries—A Symposium: Oklahoma Geological Survey Bulletin, v. 136, p. 175–191.

Ebanks, W. J., Jr., Brady, L. L., Heckel, P. H., O'Connor, H. G., Sanderson, G. A., West, R. R., and Wilson, F. W., 1979, The Mississippian and Pennsylvanian (Carboniferous) systems in the United States—Kansas: U.S. Geological Survey Professional Paper 1110-Q, 30 p.

Eble, C. F., 1994, Pallynostratigraphy of selected Middle Pennsylvanian coal beds in the Appalachian Basin, *in* Rice, C. L., ed., Elements of Pennsylvanian stratigraphy, Central Appalachian Basin: Geological Society of America Special Paper 294, p. 56–68.

Eble, C. F., and Gillespie, W. H., 1986a, Palynological studies of the Upper Kanawha Formation (Pottsville, Pennsylvanian) in West Virginia: Compass, v. 63, p. 58–65.

Eble, C. F., and Gillespie, W. H., 1986b, Characteristic small spores of the Coalburg coal (Upper Pottsville, Pennsylvanian) in West Virginia: West Virginia Academy of Science, Proceedings, v. 56, p. 104–123.

Eble, C. F., and Gillespie, W. H., 1989, Palynology of selected Pennsylvanian coal beds from the central and southern Appalachian Basin: Correlation and stratigraphic implications, *in* Englund, K. J., ed., Coal and hydrocarbon resources of North America; Volume 2, Characteristics of the Mid-Carboniferous boundary and associated coal-bearing rocks in the Appalachian Basin (28th International Geological Congress, Field Trip Guidebook T352): Washington, D.C., American Geophysical Union, p. 61–66.

Eble, C. F., and Grady, W. C., 1990, Paleoecological interpretations of a Middle Pennsylvanian coal bed in the central Appalachian basin, U.S.A.: International Journal of Coal Geology, v. 16, p. 255–286.

Eble, C. F., Gillespie, W. H., Crawford, T. J., and Rheams, L. J., 1985, Microspores in Pennsylvanian coal beds of the southern Appalachian Basin and their stratigraphic implications, *in* Englund, K. J., and others, eds., Characteristics of the Mississippian-Pennsylvanian boundary and associated coal-bearing rocks in the southern Appalachian: U.S. Geological Survey Open-File Report 85-577, p. 19–25.

Eble, C. F., Grady, W. C., and Gillespie, W. H., 1989, Palynology, petrography and paleoecology of the Hernshaw–Fire Clay coal bed in the central Appalachian Basin, *in* Cecil, C. B., and Eble, C., eds., Carboniferous geology of the eastern United States (28th International Geological Congress, Field Trip Guidebook T143): Washington, D.C., American Geophysical Union, p. 133–142.

Eggert, D. L., Chou, C.-L., Maples, C. G., Peppers, R. A., Phillips, T. L., and Rexroad, C. B., 1983, Origin and economic geology of the Springfield Coal Member in the Illinois Basin, *in* Shaver, R. H., and Sunderman, J. A., eds., Field trips in Midwestern geology (Geological Society of America Annual Meeting Guidebook): Bloomington, Indiana Geological Survey, p. 121–146.

Einor, O. L., and 14 others, 1979, The Lower-Middle Carboniferous boundary in the U.S.S.R., *in* Wagner, R. H., Higgins, A. C., and Meyen, S. V., eds., The Carboniferous of the U.S.S.R., reports presented to the I.U.G.S. Subcommission on the Carboniferous Stratigraphy at the Eighth International Congress of Carboniferous Stratigraphy and Geology, Moscow, 1975: Yorkshire Geological Society Occasional Publication 4, p. 61–81.

Englund, K. J., 1961, Regional relation of the Lee Formation to overlying formations in southeastern Kentucky and adjacent areas of Tennessee, *in* Abstracts with Programs: Geological Society of America Special Paper 68, p. 69–70.

Englund, K. J., 1979, The Mississippian and Pennsylvanian (Carboniferous) systems in the United States—Virginia: U.S. Geological Survey Professional Paper 1110-C, 21 p.

Englund, K. J., Arndt, H. H., and Henry, T. W., eds., 1979, Proposed Pennsylvanian System stratotype, Virginia and West Virginia: Ninth International Congress of Carboniferous Stratigraphy and Geology, Field Trip No. 1: American Geological Institute Guidebook Series 1, 138 p.

Englund, K. J., Arndt, H. H., Schweinfurth, S. P., Stanley, P., and Gillespie, W. H., 1986, Pennsylvanian System stratotype sections, West Virginia, *in* Neathery, T. L., ed.,: Geological Society of America, Southeastern Section, Centennial field guide: Boulder, Colorado, Geological Society of America, p. 59–68.

Ettensohn, F. R., and Peppers, R. A., 1979, Palynology and biostratigraphy of Pennington shales and coals (Chesterian) at selected sites in northeastern Kentucky: Journal of Paleontology, v. 53, p. 453–474.

Fay, R. O., Friedman, S. A., Johnson, K. S., Roberts, J. F., Rose, W. D., and Sutherland, P. K., 1979, The Mississippian and Pennsylvanian (Carboniferous) systems in the United States—Oklahoma: U.S. Geological Survey Professional Paper 1110-R, 35 p.

Felix, G. J., and Burbridge, P. P., 1967, Palynology of the Springer Formation of southern Oklahoma, USA: Palaeontology, v. 10, p. 349–425.

Fissunenko, O. P., 1974, Analog of the Cantabrian in the Carboniferous section of the Donets Basin: Akademii Nauk SSSR Izvestya, Seriya Geologicheskaya, no. 7, p. 152–166.

Fissunenko, O. P., and Laveine, J. P., 1984, Comparaison centre la distribution

des principales espèces-guides végétales du Carbonifère moyen dans le bassin du Donetz (URSS) et les bassins du Nord-Pas-de-Calais et de Lorraine (France), *in* Gordon, M., Jr., ed., Official reports: Compte Rendu, Neuvième Congrès International Stratigraphie et de Géologie du Carbonifère, Washington and Champaign-Urbana, 1979, v. 1, p. 95–106.

Furnish, W. M., and Knapp, W. D., 1966, Lower Pennsylvanian fauna from eastern Kentucky: Part 1, Ammonoids: Journal of Paleontology, v. 40, no. 2, p. 296–308.

Gao, L., 1984, Carboniferous spore assemblages in China, *in* Sutherland, P. K., and Manger, W. L., eds., Biostratigraphy: Compte Rendu, Neuvième Congrès International de Stratigraphie et de Géologie du Carbonifère, Washington and Champaign-Urbana, 1979, v. 2, p. 103–108.

Gastaldo, R. A., 1977, A middle Pennsylvanian nodule flora from Carterville, Illinois, *in* Romans, R. C., ed., Geobotany 1: New York, Plenum Press, p. 133–155.

George, T. N., and Wagner, R. H., 1972, IUGS Subcommission on Carboniferous stratigraphy, *in* Josten, K. H., ed., Proceedings and report on the General Assembly at Krefeld, August 21–22: Compte Rendu, Septième Congrès International de Stratigraphie et de Géologie du Carbonifère, Krefeld, 1971, v. 1, p. 139–147.

Gillespie, W. H., and Pfefferkorn, H. W., 1979, Distribution of commonly occurring plant megafossils in the proposed Pennsylvanian System stratotype, *in* Englund, K. J., Arndt, H. H., and Henry, T. W., eds., Proposed Pennsylvanian System stratotype, Virginia and West Virginia: Ninth International Congress Carboniferous Stratigraphy and Geology, Field Trip No. 1: American Geological Institute Selected Guidebook Series no. 1, p. 87–94.

Glenn, L. C., 1912, A geological reconnaissance of the Tradewater River region, with special reference to the coal beds: Kentucky Geological Survey, Series 3, Bulletin 17, 75 p.

Glenn, L. C., 1925, The northern Tennessee coal field: Tennessee Division of Geology Bulletin 33-B, 478 p.

Goldhammer, R. K., Oswald, E. J., and Dunn, P. A., 1991, Hierarchy of stratigraphic forcing: Example from Middle Pennsylvanian shelf carbonates of the Paradox basin, *in* Franseen, E. K., and others, eds., Sedimentary modeling: Computer simulations and methods for improved parameter definition: Kansas Geological Survey Bulletin 233, p. 361–413.

Gordon, M., Jr., 1984, Discussion following symposium, March 29, 1982, *in* Sutherland, P. K., and Manger, W. L., eds., The Atokan Series (Pennsylvanian) and its boundaries—A symposium: Oklahoma Geological Survey Bulletin 136, p. 193–198.

Grady, W. C., 1983, The petrography of West Virginia coals as an indicator of paleoclimate and coal quality: Geological Society of America Abstracts with Programs, v. 15, p. 584.

Gray, H. H., 1979, The Mississippian and Pennsylvanian (Carboniferous) systems in the United States—Indiana. U.S. Geological Survey Professional Paper 1110-K, 20 p.

Gray, L. R., 1967, Palynology of four Allegheny coals, northern Appalachian coal field: Palaeontographica B, v. 121, p. 65–86.

Grayson, R. C., Jr., 1984, Morrowan and Atokan (Pennsylvanian) conodonts from the northeastern margin of the Arbuckle Mountains southern Oklahoma, *in* Sutherland, P. K., and Manger, W. L., eds., The Atokan Series (Pennsylvanian) and its boundaries—A symposium: Oklahoma Geological Survey Bulletin 136, p. 41–54.

Grayson, R. C., Jr., and Sutherland, P. K., 1977, Conodont evidence for unconformity with Trace Creek Shale Member of the Bloyd Formation (Lower Pennsylvanian) in northwestern Arkansas and northeastern Oklahoma, *in* Sutherland, P. K., and Manger, W. L., eds., Upper Chesterian–Morrowan stratigraphy and the Mississippian-Pennsylvanian boundary in northeastern Oklahoma and northwestern Arkansas: Oklahoma Geological Survey Guidebook 18, p. 181–185.

Grayson, R. C., Jr., Merrill, G. K., Lambert, L. L., and Turner, J., 1989, Phylogenetic basis for species recognition within the conodont genus *Idiognathodus:* applicability to correlation and boundary placement, *in* Boardman, D. R., II, Barrick, J. E., Cocke, J., and Nestell, M. K., eds.,

Middle and Late Pennsylvanian chronostratigraphic boundaries in north-central Texas: Glacial-eustatic events, biostratigraphy and paleoecology: Texas Technical University Studies in Geology, v. 2, p. 75–94.

Greb, S. F., Williams, D. A., and Williamson, A. D., 1992, Geology and stratigraphy of the western Kentucky coal field: Kentucky Geological Survey Bulletin 2, series XI, 77 p.

Grebe, H., 1972, Die Verbreitung der Mikrosporen im Ruhrkarbon von den Bochumer Schichten bis zu den Dorstener Schichten (Westfal A-C): Palaeontographica B, v. 140, p. 27–115.

Greene, F. C., and Searight, W. V., 1949, Revision of the classification of the post-Cherokee beds of Missouri: Missouri Geological Survey and Water Resources Report of Investigations 11, p. 1–21.

Grosse, C. W., 1979, Miospores associated with the flint clay parting within the Fire Clay coal of the Breathitt Formation at selected sites in eastern Kentucky [M.S. thesis]: Eastern Kentucky University, Richmond, 100 p.

Grosse, C. W., and Helfrich, C. T., 1984, Miospores associated with the flint clay ("altered volcanic ash") within the Fire Clay coal of eastern Kentucky: Geological Society of America Abstracts with Programs, v. 16, p. 142.

Groves, J. R., and Grayson, R. C., Jr., 1984, Calcareous foraminifera and conodonts from the Wapanucka Formation (Lower-Middle Pennsylvanian), Frontal Ouachita Mountains, southeastern Oklahoma, in Sutherland, P. K., and Manger, W. L., eds., The Atokan Series (Pennsylvanian) and its boundaries—A symposium: Oklahoma Geological Survey Bulletin 136, p. 81–89.

Grubbs, R. K., 1984, Conodont platform elements from the Wapanucka and Atoka formations (Morrowan-Atokan) of the Mill Creek Syncline, central Arbuckle Mountains, Oklahoma, in Sutherland, P. K., and Manger, W. L., eds., The Atokan Series (Pennsylvanian) and its boundaries—A symposium: Oklahoma Geological Survey Bulletin 136, p. 65–73.

Guennel, G. K., 1952, Fossil spores of the Alleghenian coals of Indiana: Indiana Geological Survey Report of Progress 4, 40 p.

Guennel, G. K., 1958, Miospore analysis of the Pottsville coals of Indiana: Indiana Geological Survey Bulletin 13, 101 p.

Gupta, S., and Boozer, O. W., 1969, Spores and pollen from the Rock Lake Shale at Garnett locality of Kansas: J. Sen Memorial Volume, Botanical Society of Bengal, p. 69–91.

Habib, D., 1966, Distribution of spore and pollen assemblages in the Lower Kittanning Coal of western Pennsylvania: Palaeontology, v. 9, p. 629–666.

Hacquebard, P. A., 1957, Plant spores in coal from the Horton Group (Mississippian) of Nova Scotia: Micropaleontology, v. 3, p. 301–324.

Hacquebard, P. A., and Barss, M. S., 1957, A Carboniferous spore assemblage, in coal from the South Nahanni River area, Northwest Territories: Geological Survey of Canada Bulletin 40, 63 p.

Hacquebard, P. A., Barss, M. S., and Donaldson, J. R., 1960, Distribution and stratigraphic significance of small spore genera in the Upper Carboniferous of the Maritime Provinces of Canada: Compte Rendu, Quatrième Congrès International de Stratigraphie et de Géologie du Carbonifère, Heerlen, 1958, v. 1, p. 237–245.

Haley, B. R., Glick, E. E., Caplan, W. M., Holbrook, D. F., and Stone, C. G., 1979, The Mississippian and Pennsylvanian (Carboniferous) systems in the United States—Arkansas: U.S. Geological Survey Professional Paper 1110-O, 14 p.

Haq, B. U., Hardenbol, J., and Vail, P. R., 1987, Chronology of fluctuating sea levels since the Triassic: Science, v. 235, p. 1156-1167.

Harland, W. B., Armstrong, R. L., Cox, A. V., Craig, L., Smith, A. G., and Smith, D. G., 1990, A geologic time scale, 1989: Cambridge, Cambridge University Press, 263 p.

Harvey, R. D., and Dillon, J. W., 1985, Maceral distribution in Illinois coals and their paleoenvironmental implications: International Journal of Coal Geology, v. 5, p. 141–166.

Heckel, P. H., 1984, Factors in Mid-Continent Pennsylvanian limestone deposition, in Hyne, N. J., ed., Limestones of the Mid-Continent: Tulsa Geological Society Special Publication 2, p. 25–50.

Heckel, P. H., 1986, Sea-level curve for Pennsylvanian eustatic marine transgressive-regressive depositional cycles along midcontinent outcrop belt, North America: Geology, v. 14, p. 330–334.

Heckel, P. H., 1989, Current view of Midcontinent Pennsylvanian cyclothems, in Boardman, D. R., II, Barrick, J. E., Cocke, J., and Nestell, M. K., eds., Middle and Late Pennsylvanian chronostratigraphic boundaries in north-central Texas: Glacial-eustatic events, biostratigraphy and paleoecology: Part 2, Contributed papers: Texas Technical University Studies in Geology, v. 2, p. 17–34.

Heckel, P. H., 1991, Lost Branch Formation and revision of upper Desmoinesian stratigraphy along Midcontinent outcrop belt: Kansas Geological Survey, Geology Series 4, 67 p.

Helby, R., 1966, Sporologische Untersuchungen an der Karbon/Perm Grenze im Pfälzer Bergland: Fortschritte in der Geologie von Rheinland und Westfalen, 13, p. 645–704.

Helfrich, C. T., 1981, Preliminary correlations of coals of the Princess Reserve District in eastern Kentucky, in Cobb, J. C., Chesnut, D. R., Jr., Hester, N. C., and Hower, J. C., eds., Coal and coal-bearing rocks of eastern Kentucky (Geological Society of America Coal Division Field Trip): Lexington, Kentucky Geological Survey, p. 106–119.

Helfrich, C. T., 1984, Distribution of the miospore *Radiizonates* in the Pennsylvanian coals of the Princess and Licking River districts in eastern Kentucky: Geological Society of America Abstracts with Programs, v. 16, p. 145.

Henbest, L. G., 1962a, Type sections for the Morrow Series of Pennsylvanian age, and adjacent beds, Washington County, Arkansas: U.S. Geological Survey Professional Paper 450-D p. 38–41.

Henbest, L. G., 1962b, New members of the Bloyd Formation of Pennsylvanian age, Washington County, Arkansas: U.S. Geological Survey Professional Paper 450-D, p. 42–44.

Hendricks, T. A., and Read, C. B., 1934, Correlations of Pennsylvanian strata in Arkansas and Oklahoma coal fields: American Association of Petroleum Geologists Bulletin, v. 18, p. 1050–1058.

Henry, T. W., and Gordon, M., Jr., 1979, Late Devonian through early Permian (?) invertebrate faunas in proposed Pennsylvanian System stratotype area, in Englund, K. J., Arndt, H. H., and Henry, T. W., eds., Proposed Pennsylvanian System stratotype, Virginia and West Virginia: Ninth International Congress of Carboniferous Stratigraphy and Geology, Field Trip No. 1; American Geological Institute Selected Guidebook Series no. 1, p. 97–103.

Henry, T. W., Lyons, P. C., and Wendolph, J. F., Jr., 1979, Upper Pennsylvanian and Lower Permian (?) Series in the area of the proposed Pennsylvanian System stratotype, in Englund, K. J., Arndt, H. H., and Henry, T. W., eds., Proposed Pennsylvanian System stratotype, Virginia and West Virginia: Ninth International Congress of Carboniferous Stratigraphy and Geology, Field Trip No. 1; American Geological Institute Selected Guidebook Series no. 1, p. 81–85.

Hess, J. H., and Lippolt, H. J., 1986, ^{40}Ar/^{39}Ar ages of tonstein and tuff sanidines: new calibration points for the improvement of the Upper Carboniferous time scale: Chemical Geology (Isotope Geoscience Section) v. 59, p. 143–154.

Hoffmeister, W. S., Staplin, F. L., and Malloy, R. E., 1955, Mississippian plant spores from the Hardinsburg Formation of Illinois and Kentucky: Journal of Paleontology, v. 29, p. 372–399.

Hopkins, M. E., and Simon, J. A., 1975, Pennsylvanian System, in Willman, H. B., Atherton, E., Buschbach, T. C., Collinson, C., Hopkins, M. E., Lineback, J. A., and Simon, J. A., eds., Handbook of Illinois stratigraphy: Illinois State Geological Survey Bulletin 95, p. 163–201.

Horowitz, A. S., 1979, The Mazon Creek Flora: review of research and bibliography, in Nitecki, N. H., ed., Mazon Creek Fossils: New York, Academic Press, p. 143–158.

Howard, R. H., 1979a, Carboniferous cyclicity related to the development of the Mississippian-Pennsylvanian unconformity in the Illinois Basin, in Timofeyev, P. P., ed., Coalbearing formations: Compte Rendu, Huitième Congrès International de Stratigraphie et de Géologie du Carbonifère, Moscow, 1975, v. 5, p. 32–41.

Howard, R. H., 1979b, The Mississippian-Pennsylvanian unconformity in the

Illinois Basin—old and new thinking, *in* Palmer, J. E., and Dutcher, R. R., eds., Depositional and structural history of the Pennsylvanian System of the Illinois Basin: Ninth International Congress of Carboniferous Stratigraphy and Geology, Champaign-Urbana, Field Trip 9, part 2; Illinois State Geological Survey Guidebook 15a, p. 34–43.

Howe, W. B., 1953, Upper Marmaton strata in western and northern Missouri: Missouri Geological Survey and Water Resources Report of Investigations no. 9, 29 p.

Howe, W. B., 1982, Stratigraphy of the Pleasanton Group, Pennsylvanian System in Missouri: Missouri Department of Natural Resources Geology and Land Survey Division Open File Report Series 82-10-GI, 77 p.

Howe, W. B., and Koenig, J. W., 1961, The stratigraphic succession in Missouri: Missouri Geological Survey and Water Resources, series 2, v. 40, 185 p.

Hower, J. C., Fiene, F. L., Wild, G. D., and Helfrich, C. T., 1983, Coal metamorphism in the upper portion of the Pennsylvanian Sturgis Formation in western Kentucky, Geological Society of America Bulletin, v. 94, p. 1475–1481.

Hower, J. C., Rathibone, R. F., and Eble, C. F., 1992, No. 5 Block coal bed, northeastern Kentucky: Geological Society of America Abstracts with Programs, v. 24, no. 3, p. 29.

Hower, J. C., Eble, C. F., and Rathbone, R. F., 1994, Petrology and palynology of the No. 5 Block coal bed, northeastern Kentucky: International Journal of Coal Geology, v. 25, p. 171–193.

Ibrahim, A. C., 1933, Sporenformen des Ägirhorizontes des Ruhr-Reviers: Wurzburg, Konrad Triltsch, 47 p.

Inosova, K. I., Kruzina, A. K., and Shvartsman, E. G., 1975, Fig. 6, The scheme of distribution of characteristic genera and species of microspores and pollen in Carboniferous and Lower Permian deposits, *in* Aizenverg, D. E., Lagutina, V. V., Levenshtein, M. L., and Popov, V. S., eds., Field excursion guidebook for the Donets Basin: Eighth International Congress of Carboniferous Stratigraphy and Geology, Moscow: Ministry of Geology of Ukraine, Nauka, p. 287.

Inosova, K. I., Kruzina, A. K., and Shvartsman, E. G., 1976, Atlas of the microspores and pollen of the Upper Carboniferous and Lower Permian of the Donets Basin: Moscow, Geologic Ministry, 154 p.

Jacobson, R. J., 1992, Geology of the Goreville Quadrangle, Johnson and Williamson counties, Illinois: Illinois State Geological Survey Bulletin 97, 32 p.

Jacobson, R. J., Trask, C. B., and Norby, R. D., 1983, A Morrowan/Atokan limestone from the lower Abbott Formation of southern Illinois: Geological Society of America Abstracts with Programs, v. 15, p. 602–603.

Jacobson, R. J., Trask, C. B., Ault, C. H., Carr, D. D., Gray, H. H., Hasenmueller, W. A., Williams, D., and Williamson, A. D., 1985, Unifying nomenclature in the Pennsylvanian System of the Illinois Basin: Illinois State Academy of Science Transactions, v. 78, p. 1–11.

Janssen, R. E., 1939, Leaves and stems from fossil forests: Illinois State Museum Popular Science Series 1, 190 p.

Jennings, J. R., 1981, Pennsylvanian plants of eastern Kentucky: Compression fossils from the Breathitt Formation near Hazard, Kentucky, *in* Cobb, J. C., Chesnut, D. R., Jr., Hester, N. C., and Hower, J. C., eds., Coal and coal-bearing rocks in eastern Kentucky (Geological Society of America Coal Division Field Trip): Lexington, Kentucky Geological Survey, p. 147–162.

Jennings, J. R., 1984, Distribution of fossil plant taxa in the Upper Mississippian and Lower Pennsylvanian of the Illinois Basin, *in* Sutherland, P. K., and Manger, W. L., eds., Biostratigraphy: Compte Rendu, Neuvième Congrès International de Stratigraphie et de Géologie du Carbonifère, Washington and Champaign-Urbana, 1979, v. 2, p. 301–312.

Jennings, J. R., and Fraunfelter, G. H., 1986, Preliminary report on macropaleontology of strata above and below the upper boundary of the type Mississippian: Illinois State Academy of Science Transactions, v. 79, p. 253–261.

Jewett, J. M., O'Conner, H. G., and Zeller, D. E., 1968, Pennsylvanian System, *in* Zeller, D. E., ed., The stratigraphic succession in Kansas: Kansas Geological Survey Bulletin 189, p. 21–43.

Jongmans, W. J., and Gothan, W., 1934, Florenfolge und vergleichende Stratigraphie des Karbons der östlichen Staaten Nord-Amerikas, Vergleich mit West-Europa: Jaarverslag Geologisch Bureau voor het Nederlandsche Mijngebied te Heerlen, p. 17–44.

Jongmans, W. J., and Gothan, W., 1937, Betrachtungen über die Ergebnisse des zeweiten Kongresses für Karbonstratigraphie: Compte Rendu, Deuxième Congrès pour l'Avancement des Etudes Stratigraphie et de Géologie du Carbonifère, Heerlen, 1935, v. 1, p. 1–40.

Kehn, T. M., 1974, Geologic map of parts of the Dekoven and Saline Mines quadrangles, Crittendon and Union counties, Kentucky: U.S. Geological Survey Geologic Quadrangle Map 1147, scale 1:24,000.

Kehn, T. M., Palmer, J. E., and Franklin, G. J., 1967, Revised correlation of the No. 4 (Dawson Springs No. 6) coal bed, western Kentucky coal field: U.S. Geological Survey Professional Paper 575-C, p. 160–164.

Keyes, C. R., 1893, Geological formations of Iowa: Iowa Geological Survey, v. 1, p. 86–140.

Klein, G. deV., 1990, Pennsylvanian time scales and cyclic periods: Geology, v. 18, p. 455–457.

Klein, G. deV., and Kupperman, J. B., 1992, Pennsylvanian cyclothems; methods of distinguishing tectonically induced changes in sea level from climatically induced changes: Geological Society of America Bulletin, v. 104, p. 166–175.

Klein, G. deV., and Willard, D. A., 1989, Origin of the Pennsylvanian coal-bearing cyclothems of North America: Geology, v. 17, p. 152–155.

Knox, E. M., 1950, The spores of *Lycopodium, Phylloglossum, Selaginella, and Isoetes* and their value in the study of microfossils of Paleozoic age: Botanical Society of Edinburgh Transactions and Proceedings, v. 35, p. 209–357.

Kosanke, R. M., 1943, The characteristic plant microfossils of the Pittsburgh and Pomeroy Coals of Ohio: American Midland Naturalist, v. 29, p. 119–132.

Kosanke, R. M., 1947, Plant microfossils in correlation of coal beds: Journal of Geology, v. 55, p. 280–284.

Kosanke, R. M., 1950, Pennsylvanian spores of Illinois and their use in correlation: Illinois State Geological Survey Bulletin 74, 128 p.

Kosanke, R. M., 1959a, Late Paleozoic small spore floras of United States: Compte Rendu, Résumés 2, Neuvième Congrès International Botanique, Montreal, 1959, p. 200.

Kosanke, R. M., 1959b, *Wilsonites*, new name for *Wilsonia* Kosanke 1950: Journal of Paleontology, v. 33, p. 700.

Kosanke, R. M., 1964, Applied Paleozoic palynology, *in* Cross, A. T., ed., Palynology in oil exploration: Society of Economic Paleontologists and Mineralogists Special Publication 11, p. 75–89.

Kosanke, R. M., 1973, Palynological studies of the coals of the Princess Reserve District in northeastern Kentucky: U.S. Geological Survey Professional Paper 839, 22 p.

Kosanke, R. M., 1981, Palynomorph content of the Hazard No. 7 coal bed, Kentucky, *in* Cobb. J. C., Chesnut, D. R., Jr., Hester, N. C., and Hower, J. C., eds., Coal and coal-bearing rocks of eastern Kentucky (Geological Society of America Coal Division Field Trip): Lexington, Kentucky Geological Survey, p. 163–165.

Kosanke, R. M., 1982, Mississippian-Pennsylvanian boundary in the United States based on palynomorphs, *in* Ramsbottom, W. H. C., Saunders, W. B., and Owens, B., eds., Biostratigraphic data for a Mid-Carboniferous boundary: Leeds, England, International Union of Geological Sciences Subcommission on Carboniferous Stratigraphy, Institute of Geological Sciences, p. 27–35.

Kosanke, R. M., 1984, Palynology of selected coal beds in the proposed Pennsylvanian System stratotype in West Virginia: U.S. Geological Survey Professional Paper 1318, 44 p.

Kosanke, R. M., 1988a, Palynological studies of Middle Pennsylvanian coal beds of the proposed Pennsylvanian System stratotype in West Virginia: U.S. Geological Survey Professional Paper 1455, 73 p.

Kosanke, R. M., 1988b, Palynological studies of Lower Pennsylvanian coal beds and adjacent strata of the proposed Pennsylvanian System stratotype in

Virginia and West Virginia: U.S. Geological Survey Professional Paper 1479, 17 p.

Kosanke, R. M., 1988c, Palynological analyses of Upper Pennsylvanian coal beds and adjacent strata from the proposed Pennsylvanian System stratotype in West Virginia: U.S. Geological Survey Professional Paper 1486, 24 p.

Kosanke, R. M., and Cecil, C. B., 1989, Late Pennsylvanian climate changes and palynomorph extinctions: International Geological Congress, 28th, Washington, D.C., 1989, Abstracts, v. 2, p. 214–215.

Kosanke, R. M., and Cecil, C. B., 1992, Pennsylvanian climatic changes as evidenced by palynomorph extinctions: Geological Society of America Abstracts with Programs, v. 24, no. 7, p. 163.

Kosanke, R. M., and Peppers, R. A., 1981, Spores from Chesterian coals of the Illinois Basin: Geological Society of America Abstracts with Programs, v. 13, p. 490.

Kosanke, R. M., Simon, J. A., Wanless, H. R., and Willman, H. B., 1960, Classification of the Pennsylvanian strata of Illinois: Illinois State Geological Survey Report of Investigations 124, 84 p.

Krebs, C. E., and Teets, D. D., Jr., 1914, Kanawha County: West Virginia Geological Survey County Report, 679 p.

Lambert, L. L., 1989, A proposed basis for establishing a formal Atokan/Desmoinesian boundary: Geological Society of America Abstracts with Programs, v. 21, no. 1, p. 16.

Lambert, L. L., 1990, Atokan conodonts from the type Des Moines region: Geological Society of America Abstracts with Programs, v. 22, no. 1, p. 12.

Lambert, L. L., 1992, Atokan and basal Desmonesian conodants from central Iowa, reference area for the Desmoinesian Stage, *in* Sutherland, P. K., and Manger, W. L., eds., Recent advances in Middle Carboniferous biostratigraphy—A symposium: Oklahoma Geological Survey Circular 94, p. 111–123.

Lambert, L. L., and Thompson, T. L., 1990, Age and significance of the "Ladden Branch" Limestone (Riverton Formation; Pennsylvanian), west-central Missouri: Geological Society of America Abstracts with Programs, v. 22, no. 1, p. 13.

Landis, E. R., and Van Eck, O. J., 1965, Coal resources of Iowa: Iowa Geological Survey Technical Paper 4, 141 p.

Lane, H. R., and Straka, J. J., II, 1974, Late Mississippian and Early Pennsylvanian conodonts, Arkansas and Oklahoma: Geological Survey of America Special Paper 152, 144 p.

Lane, H. R., and West, R. R., 1984, The inadequacies of Carboniferous Serial nomenclature: North American examples, *in* Sutherland, P. K., and Manger, W. L., eds., The Atokan Series (Pennsylvanian) and its boundaries—A symposium: Oklahoma Geological Survey Bulletin, v. 136, p. 91–100.

Lane, H. R., Merrill, G. K., Straka, J. J., II, and Webster, G. D., 1971, North American Pennsylvanian conodont biostratigraphy, *in* Sweet, W. C., and Bergstrom, S. M., eds., Symposium on conodont biostratigraphy: Geological Society of America Memoir 127, p. 395–414.

Langenheim, R. L., Jr., 1991, Pennsylvanian time scales and cyclic period: Comment: Geology, v. 19, p. 405.

Langenheim, R. L., Jr., and Scheihing, M. H., 1983, Detailed biostratigraphic correlation, Shumway "Cyclothem," Virgilian, Illinois Basin: Geological Society of America, Abstracts with Programs, v. 15, p. 265.

Langford, G., 1958, The Wilmington coal flora from a Pennsylvanian deposit in Will County, Illinois: Downers Grove, Illinois, Esconi Associates, 360 p.

Langford, G., 1963, The Wilmington coal fauna and additions to Wilmington coal flora from a Pennsylvanian deposit in Will County, Illinois: Downers Grove, Illinois, Esconi Associates, 280 p.

Laveine, J. P., 1976, Report on the Westphalian D: *in* Holub, V. M., and Wagner, R. H., eds., Symposium on Carboniferous Stratigraphy: I.U.G.S. Subcommission on Carboniferous Stratigraphy, 1973, Prague, p. 71–87.

Leary, R. L., 1974, Stratigraphy and floral characteristics of the basal Pennsylvanian strata in west-central Illinois: Compte Rendu, Septième Congrès International de Stratigraphie et de Géologie du Carbonifère, Krefeld, Germany, 1971, v. 3, p. 341–350.

Leary, R. L., 1975, Early Pennsylvanian paleogeography of an upland area, western Illinois, U.S.A.: Bulletin de la Societé Géologique de Belgique 84, p. 19–31.

Leary, R. L., 1981, Early Pennsylvanian geology and paleobotany of the Rock Island County, Illinois, area—part 1: Geology: Illinois State Museum Report of Investigations, v. 37, 88 p.

Leary, R. L., 1984, Topography and geology of the Early Pennsylvanian erosional surface on the northwest margin of the Illinois Basin, USA, *in* Sutherland, P. K., and Manger, W. L., eds., Biostratigraphy: Compte Rendu, Neuvième Congrès International de Stratigraphie et de Géologie du Carbonifère, Washington and Champaign-Urbana, 1979, v. 2, p. 391–398.

Leary, R. L., 1985, Early Pennsylvanian paleotopography and depositional environments, Rock Island County, Illinois: Illinois State Geological Survey Guidebook 18, 42 p.

Leary, R. L., and Pfefferkorn, H. W., 1977, An early Pennsylvanian flora with *Megalopteris* and Noeggerathiales from west-central Illinois: Illinois State Geological Survey Circular 500, 77 p.

Lee, W., 1916, Geology of the Shawneetown quadrangle in Kentucky: Kentucky Geological Survey, series 4, v. 4, part 2, 73 p.

Lesley, J. P., 1876, The Boyd's Hill gas well at Pittsburgh: Second Pennsylvania Geological Survey, v. L, p. 217–237.

Lesquereux, L., 1857, Palaeontological report of the fossil flora of the coal measures of the western Kentucky coal field, *in* Owen, D. D., ed., Third report of the Geological Survey of Kentucky: Lexington, Kentucky, p. 501–576.

Lesquereux, L., 1870, Report on the fossil plants of Illinois, *in* Worthen, A. H., ed., Geology and paleontology: Illinois State Geological Survey, v. 4, p. 375–508.

Li, X., and Yao, Z., 1982, A review of recent research on the Cathaysia flora in Asia: American Journal of Botany, v. 69, p. 479–486.

Li, X., and Zhang, L., 1983, The Upper Carboniferous of China, *in* Martinez Diaz, C., ed., The Carboniferous of the World, China, Korea, Japan, and S.E. Asia: International Union of Geological Sciences Publication 16, p. 87–121.

Liabeuf, J. J., and Alpern, B., 1969, Étude palynologique du bassin houiller de St. Etienne, stratotype du Stéphanien: Compte Rendu, Sixième Congrès International de Stratigraphie et de Géologie du Carbonifère, Sheffield, 1967, v. 1, p. 155–169.

Liabeuf, J. J., and Alpern, B., 1970, Le gisement houiller de Decize Etude palynologique: Compte Rendu, Sixième Congrès International de Stratigraphie et de Géologie du Carbonifère, Sheffield, 1967, v. 3, p. 1083–1100.

Liabeuf, J. J., Doubinger, J., and Alpern, B., 1967, Caractères palynologiques des charbons du Stéphanien de quelques gisements français: Revue de Micropaléontologie, v. 10, p. 3–14.

Loboziak, S., 1971, Les micro- et mégaspores de la partie occidentale du bassin houiller du Nord de la France: Palaeontographica B, v. 132, p. 1–127.

Loboziak, S., 1972, Une microflore d'âge Namurien ou Westphalien inférieur de la Carrière Napoléon, Ferques (Pas-de-Calais), France: Review of Palaeobotany and Palynology, v. 13, p. 125–146.

Loboziak, S., 1974, Considération palynologiques sur le Westphalien d'Europe occidentale: Review of Palaeobotany and Palynology, v. 18, p. 271–289.

Loboziak, S., Coquel, R., and Owens, B., 1984, Les microspores des Formations Hale et Bloyd du Nord de l'Arkansas, *in* Sutherland, P. K., and Manger, W. R., eds., Biostratigraphy: Compte Rendu, Neuvième Congrès International de Stratigraphie et de Géologie Carbonifère, Washington and Champaign-Urbana, 1979, v. 2, p. 385–390.

Loose, F., 1934, Sporenformen aus dem Flöz Bismarck des Ruhrgebietes: Arbeiten aus dem Institut fur Paläobotanik und Petrographie der Brennsteine, Berlin, v. 4, p. 127–164.

Love, L. G., 1960, Assemblages of small spores from the Lower Oil-Shale Group of Scotland: Royal Society of Edinburgh Proceedings, sec. B, v. 67, p. 99–126.

Lyons, P. C., 1984, Carboniferous megafloral zonation of New England, *in* Sutherland, P. K., and Manger, W. L., eds., Biostratigraphy: Compte

Rendu, Neuvième Congrès International de Stratigraphie et de Géologie du Carbonifère, Washington and Champaign-Urbana, 1979, v. 2, p. 483–502.

Lyons, P. C., and Alpern, B., eds., 1989, Peat and coal: Origin, facies, and depositional models: International Journal of Coal Geology, v. 12, 798 p.

Lyons, P. C., Callcott, T. G., and Alpern, B., eds., 1990, Peat and coal: origin, facies, and coalification: International Journal of Coal Geology, v. 16, 237 p.

Mahaffy, J. F., 1985, Profile patterns of coal and peat palynology in the Herrin (No. 6) Coal Member, Carbondale Formation, Middle Pennsylvanian of southern Illinois, *in* Dutro, J. T., and Pfefferkorn, H. W., eds., Paleontology, paleoecology, and paleogeography: Compte Rendu, Neuvième Congrès International de Stratigraphie et de Géologie du Carbonifère, Washington and Champaign-Urbana, 1979, v. 5, p. 25–34.

Manger, W. L., and Sutherland, P. K., 1984, The Mississippian-Pennsylvanian boundary in the southern Midcontinent, United States, *in* Sutherland, P. K., and Manger, W. L., eds., Biostratigraphy: Compte Rendu, Neuvième Congrès International de Stratigraphie et de Géologie du Carbonifère, Washington and Champaign-Urbana, 1979, v. 2, p. 369–376.

Manos, C. T., 1963, Petrography and depositional environment of the Sparland Cyclothem [Ph.D. thesis]: Urbana, University of Illinois, 110 p.

Marshall, A. E., and Smith, A. H. V., 1965, Assemblages of miospores from some Upper Carboniferous coals and their associated sediments in the Yorkshire coalfield: Palaeontology, v. 7, p. 656–673.

Maynard, J. P., and Leeder, M. R., 1992, On the periodicity and magnitude of Late Carboniferous glacio-eustatic sea level changes: Geological Society of London Journal, v. 149, p. 303–311.

McCabe, L. C., 1932, Some plant structures of coal: Illinois State Academy of Science Transactions, 1931, v. 24, p. 321-326.

McCabe, P. J., 1991, Geology of coal; Environments of deposition, *in* Gluskoter, H. J., Rice, D. D., and Taylor, R. B., eds., Economic geology, U.S.: Boulder, Colorado, Geological Society of America, Geology of North America, v. P-2, p. 469–482.

McDowell, R. C., Grabowski, G. J., Jr., and Moore, S. L., 1981, Geologic map of Kentucky: U.S. Geological Survey, 4 sheets, scale 1:250,000.

McGregor, D. C., 1973, Lower and Middle Devonian spores of eastern Gaspe, Canada, 1: Systematics: Palaeontographica B, v. 142, p. 1–77.

McKee, E. D., and Crosby, E. J., 1975, Paleotectonic investigations of the Pennsylvanian System in the United States, part I. Introduction and regional analysis of the Pennsylvanian System: U.S. Geological Survey Professional Paper 853, 349 p.

Merrill, G. K., 1964, Zonation of platform conodont genera in Conemaugh strata of Ohio and vicinity [M.S. thesis]: Austin, University of Texas, 175 p.

Milici, R. C., Briggs, G., Knox, L. M., Sitterly, P. D., and Statler, A. T., 1979, The Mississippian and Pennsylvanian (Carboniferous) systems in the United States—Tennessee: U.S. Geological Survey Professional Paper 1110-G, 38 p.

Miller, M. A., Eames, L. E., and Prezbindowski, D. R., 1989, Upper Mississippian and Lower Pennsylvanian lithofacies and palynology from northeastern Oklahoma: A field excursion: American Association of Stratigraphic Palynologists, Field Trip Guidebook, p. 2–61.

Miller, M. S., 1974, Stratigraphy and coal beds of Upper Mississippian and Lower Pennsylvanian rocks in southwestern Virginia: Virginia Division of Mineral Resources Bulletin 84, 211 p.

Moore, R. C., 1931, Pennsylvanian cycles in the northern Midcontinent region: Illinois State Geological Survey Bulletin 60, p. 247–257.

Moore, R. C., 1932, A reclassification of the Pennsylvanian System in the northern Midcontinent region: Kansas Geological Society, Sixth Annual Field Conference Guidebook, p. 79–98.

Moore, R. C., coordinator, 1944, Correlation of Pennsylvanian formations of North America: Geological Society of America Bulletin, v. 55, p. 657–706.

Moore, R. C., 1947, The Morrowan series of northeastern Oklahoma: Oklahoma Geological Survey Bulletin 66, 151 p.

Moore, R. C., 1948, Classification of Pennsylvanian rocks in Iowa, Kansas, Missouri, Nebraska, and northern Oklahoma: American Association of Petroleum Geologists Bulletin, v. 32, p. 2011–2040.

Moore, R. C., and Thompson, M. L., 1949, Main divisions of Pennsylvanian Period and System: American Association of Petroleum Geologists Bulletin, v. 33, p. 275–301.

Morgan, J. L., 1955a, The correlation of certain Desmoinsian coal beds of Oklahoma by spores [M.S. thesis]: Norman, The University of Oklahoma, 118 p.

Morgan, J. L., 1955b, Spores of McAlester coal: Oklahoma Geological Survey Circular 36, 52 p.

Neal, D. W., 1973, Palynology of the coals of the lower tongue of the Breathitt Formation (Lower Pennsylvanian) of eastern Kentucky [M.S. thesis]: Richmond, Eastern Kentucky University, 66 p.

Nelson, W. J., and Lumm, D. K., 1990, Geologic map of the Eddyville quadrangle, Illinois: Illinois State Geological Survey Map IGQ-5, scale 1:24,000.

Nelson, W. J., and 11 others, 1991, Geology of the Eddyville, Stonefort, and Creal Springs quadrangles, Southern Illinois: Illinois State Geological Survey Bulletin 96, 85 p.

Neves, R., 1958, Upper Carboniferous plant spore assemblages from the *Gastrioceras subcrenatum* horizon, North Staffordshire: Geological Magazine, v. 95, p. 1–19.

Neves, R., 1961, Namurian plant spores from the southern Pennines, England: Palaeontology, v. 4, p. 247–279.

Newberry, J. S., 1874, Circles of deposition in American sedimentary rocks: American Association for Advancement of Science Proceedings, v. 22, p. 185–196.

Noé, A. C., 1925, Pennsylvanian flora of northern Illinois: Illinois State Geological Survey Bulletin 52, 113 p.

Novik, E. O., and Fissunenko, O. P., 1978, On the position of the Middle-Upper Carboniferous boundary, *in* Meyen, S. V., and others, eds., General problems of the Carboniferous stratigraphy: Compte Rendu, Huitième Congrès International de Stratigraphie et de Géologie du Carbonifère, Moscow, 1975, v. 1, p. 119–126.

Ouyang, S., and Li, Z., 1980, Upper Carboniferous spores from Shuo Xian, northern Shanxi: Fifth International Palynology Conference: Nanjing, China, Nanjing Institute of Geology and Paleontology, 16 p.

Owen, D. D., 1855, Geological report to the president and stockholders of the Saline Coal and Manufacturing Company: Cincinnati, Ohio, T. Wrightson and Co., p. 25–54.

Owen, D. D., 1856, Report of the geological survey in Kentucky made during the years 1854 and 1855: Kentucky Geological Survey Bulletin, ser. 1, v. 1, 416 p.

Owens, B., 1984, Miospore zonation of the Carboniferous, *in* Sutherland, P. K., and Manger, W. L., eds., Biostratigraphy: Compte Rendu, Neuvième Congrès International de Stratigraphie et de Géologie du Carbonifère, Washington and Champaign-Urbana, 1979, v. 2, p. 90–102.

Owens, B., Neves, R., Gueinn, K. J., Mishell, D., Sabry, R. F., Hassan, S. M. Z., and Williams, J. E., 1977, Palynological division of the Namurian of northern England and Scotland: Yorkshire Geological Society Proceedings, v. 41, p. 381–398.

Owens, B., Loboziak, S., and Teteriuk, V. K., 1978, Palynological subdivision of the Dinantian to Westphalian deposits of northwest Europe and the Donetz Basin of the U.S.S.R.: Palynology, v. 2, p. 69–91.

Owens, B., Loboziak, S., and Coquel, R., 1984, Late Mississippian–Early Pennsylvanian miospore assemblages from northern Arkansas, *in* Sutherland, P. K., and Manger, W. L., eds., Biostratigraphy: Compte Rendu, Neuvième Congrès International de Stratigraphie et de Géologie du Carbonifère, Washington and Champaign-Urbana, 1979, v. 2, p. 377–384.

Owens, B., Riley, N. J., and Calver, M. A., 1985, Boundary stratotypes and new stage names for the lower and middle Westphalian sequences in Britain, *in* Cross, A. T., ed., Economic geology: coal, oil and gas: Compte Rendu, Dixième Congrès International de Stratigraphie et de Géologie du Carbonifère, Madrid, 1983, v. 4, p. 461–472.

Owens, B., Clayton, G., Gao, L., and Loboziak, S., 1989, Miospore correlation of the Carboniferous deposits of Europe and China: Compte Rendu, Onzième Congrès International de Stratigraphie et de Géologie du Carbonifère, Beijing, 1987, v. 3, p. 189–210.

Paproth, E., and 12 others, 1983, Bio- and lithostratigraphic subdivisions of the Silesian in Belgium, a review: Annals de la Société Geologique de Belgique, v. 106, p. 241–283.

Parrish, J. T., 1982, Upwelling and petroleum source bed, with reference to Paleozoic: American Association Petroleum Geologists Bulletin, v. 66, p. 750–774.

Parrish, J. T., 1993, Climate of the supercontinent Pangea: Journal of Geology, v. 101, p. 215–233.

Pearson, D. L., 1975, Palynology of the middle and upper Seminole coals (Pennsylvanian) of Tulsa County, Oklahoma [M.S. thesis]: Norman, The University of Oklahoma, 72 p.

Peppers, R. A., 1964, Spores in strata of Late Pennsylvanian cyclothems in the Illinois Basin: Illinois State Geological Survey Bulletin 90, 89 p.

Peppers, R. A., 1970, Correlation and palynology of coals in the Carbondale and Spoon formations (Pennsylvanian) of the northeastern part of the Illinois Basin: Illinois State Geological Survey Bulletin 93, 173 p.

Peppers, R. A., 1977, Palynology and correlations of some coals in the Tradewater Formation (Pennsylvanian) in western Kentucky, and in the Abbott Formation in southeastern Illinois: Geological Society of America Abstracts with Programs, v. 9, p. 640–641.

Peppers, R. A., 1979, Development of coal-forming floras during the early part of the Pennsylvanian in the Illinois Basin, *in* Palmer, J. E., and Dutcher, R. R., eds., Depositional and structural history of the Pennsylvanian System of the Illinois Basin: Invited papers: Ninth International Congress of Carboniferous Stratigraphy and Geology, Field Trip 9, part 2; Illinois State Geological Survey Guidebook Series 15a, p. 8–14.

Peppers, R. A., 1982a, Palynology of coals along Roaring Creek, *in* Eggert, D. E., and Phillips, T. L., eds., Environments of deposition—coal balls, cuticular shale and gray-shale floras in Fountain and Parke counties, Indiana: Indiana Geological Survey Special Report 30, p. 14–19.

Peppers, R. A., 1982b, Palynology of the unnamed coal in the Staunton Formation, Maple Grove Mine, *in* Eggert, D. L., and Phillips, T. L., eds., Environments of depositions—coal balls, cuticular shale and gray-shale floras in Fountain and Parke counties, Indiana: Indiana Geological Survey Special Report 30, p. 27–33.

Peppers, R. A., 1984, Comparison of miospore assemblages in the Pennsylvanian System of the Illinois Basin with those in the Upper Carboniferous of western Europe, *in* Sutherland, P. K., and Manger, W. L., eds., Biostratigraphy: Compte Rendu, Neuviéme Congrès International de Stratigraphie et de Géologie du Carbonifère, Washington and Champaign-Urbana, 1979, v. 2, p. 483–502.

Peppers, R. A., 1986, Palynological changes at the Desmoinesian-Missourian (Pennsylvanian) boundary and some possible causes: Geological Society of America Abstracts with Programs, v. 18, p. 319.

Peppers, R. A., 1993, Correlation of the "Boskydell Sandstone" and other sandstones containing marine fossils in southern Illinois using palynology of adjacent coal beds: Illinois State Geological Survey Circular 553, 18 p.

Peppers, R. A., 1994, Palynology of the Desmoinesian-Missourian transition in the Lost Branch Formation of Kansas: Geological Society of America Abstracts with Programs, v. 26, no. 5, p. 57.

Peppers, R., A., 1995, Palynological correlation of the Lewisport Coal Bed (early Desmoinesian) and equivalent coals in the Illinois Basin: Illinois Basin studies: Illinois Basin Consortium of Kentucky, Indiana, and Illinois State Geological Surveys, in press.

Peppers, R. A., and Pfefferkorn, H. W., 1970, A comparison of the floras of the Colchester (No. 2) Coal and Francis Creek Shale, *in* Smith, W. H., Nance, R. B., Hopkins, M. E., Johnson, R. G., and Shabica, C. W., eds., Depositional environments in parts of the Carbondale Formation: Geological Society of America Coal Division field trip, Illinois State Geological Survey Guidebook Series 8, p. 61–74.

Peppers, R. A., and Phillips, T. L., 1972, Pennsylvanian coal-swamp floras in the Illinois Basin: Geological Society of America Abstracts with Programs, v. 4, p. 624–625.

Peppers, R. A., and Popp, J. T., 1979, Stratigraphy of the lower part of the Pennsylvanian System in southeastern Illinois and adjacent portions of Indiana and Kentucky, *in* Palmer, J. E., and Dutcher, R. R., eds., Depositional and structural history of the Pennsylvanian System: Part 2: Invited papers: Ninth International Congress of Carboniferous Stratigraphy and Geology, Champaign-Urbana, Field Trip 9, part 2: Illinois State Geological Survey Guidebook Series 15a, p. 65–72.

Peppers, R. A., Howe, W. B., and Deason, K., 1993, Palynological zonation and physical stratigraphy of pre-Desmoinesian strata along a subsurface cross section in northwestern Missouri: Geological Society of America Abstracts with Programs, v. 25, no. 3, p. 72.

Pfefferkorn, H. W., 1979, High diversity and stratigraphic age of the Mazon Creek Flora, *in* Nitecki, M. H., ed., Mazon Creek fossils: New York, Academic Press, p. 129–142.

Pfefferkorn, H. W., and Gillesple, W. H., 1980, Biostratigraphy and biogeography of plant compression fossils in the Pennsylvanian of North America, *in* Dutcher, D. L., and Taylor, T. N., eds., Biostratigraphy of fossil plants, successional and paleoecological analyses: Stroudsburg, Pennsylvania, Dowden, Hutchinson, and Ross, p. 93–118.

Pfefferkorn, H. W., and Thomson, M. C., 1982, Changes in dominance patterns in Upper Carboniferous plant-fossil assemblages: Geology, v. 10, p. 641–644.

Pfefferkorn, H. W., Peppers, R. A., and Phillips, T. L., 1971, Some fern-like fructifications and their spores from the Mazon Creek Compression flora of Illinois (Pennsylvanian): Illinois State Geological Survey Circular 463, 55 p.

Phillips, T. L., 1981, Stratigraphic occurrences and vegetational patterns of Pennsylvanian pteridosperms in Euramerican coal swamps: Review of Palaeobotany and Palynology, v. 32 p. 5–26.

Phillips, T. L., and Cross, A. T., 1991, Paleobotany and paleoecology of coal, *in* Gluskoter, H. J., Rice, D. D., and Taylor, R. B., eds., Economic geology, U.S.: Boulder, Colorado, Geological Society of America, Geology of North America, v. P-2, p. 483–502.

Phillips, T. L., and DiMichele, W. A., 1981, Paleoecology of the Middle Pennsylvanian age coal swamps in southern Illinois/Herrin Coal Member at Sahara Mine No. 6, *in* Niklas, K. J., ed., Paleobotany, paleoecology, and evolution, Volume 1: New York, Praeger, p. 231–284.

Phillips, T. L., and DiMichele, W. A., 1992, Comparative ecology and life-history of arborescent lycopsids in Late Carboniferous swamps of Euramerica, Missouri Botanical Garden Annals, v. 79, p. 560–588.

Phillips, T. L., and Peppers, R. A., 1984, Changing patterns of Pennsylvanian coal-swamp vegetation and implications of climatic control on coal occurrence: International Journal of Coal Geology, v. 3, p. 205–255.

Phillips, T. L., Pfefferkorn, H. W., and Peppers, R. A., 1973, Development of paleobotany in the Illinois Basin: Illinois State Geological Survey Circular 480, 86 p.

Phillips, T. L., Peppers, R. A., Avcin, M. J., and Laughnan, P. F., 1974, Fossil plants and coal: Patterns of change in Pennsylvanian coal swamps of the Illinois Basin: Science, v. 184, p. 1367–1369.

Phillips, T. L., Peppers, R. A., and DiMichele, W. A., 1985, Stratigraphic and interregional changes in Pennsylvanian coal-swamp vegetation: environmental inferences: International Journal of Coal Geology, v. 5, p. 43–110.

Pierce, W. G., and Courtier, W. H., 1937, Geology and coal resources of southeastern Kansas coalfield: Kansas Geological Survey Bulletin 24, 122 p.

Pi-Radondy, M., and Doubinger, J., 1968, Spores nouvelles de Stéphanien (Massif Central français): Pollen et Spores, v. 10, p. 411–430.

Platt, F., 1875, Report of progress in the Clearfield and Jefferson district of the bituminous coal fields of western Pennsylvania: Second Pennsylvania Geological Survey, v. H, 203 p.

Playford, G., 1962–63, The Lower Carboniferous microfloras of Spitsbergen: Palaeontology, v. 5 p. 550–678.

Posamentier, H. W., and Vail, P. R., 1988, Eustatic controls on clastic deposition II—sequence and systems tract models, *in* Wilgus, C. K., Hastings,

B. S., Kendall, C. G. St., Posamentier, H. W., Ross, C. A., and Van Wagoner, J. C., eds., Sea level change: An integrated approach: Society of Economic Paleontologists and Mineralogists Special Publication 42, p. 125–154.

Posamentier, H. W., and Weimer, P., 1993, Siliciclastic sequence stratigraphy and petroleum geology—where to from here?: American Association of Petroleum Geologists Bulletin, v. 77, p. 731–742.

Posamentier, H. W., Summerhayes, C. P., Hag, B. U., and Allen, G. P., 1993, eds., Sequence stratigraphy and facies associations: International Association of Sedimentologists Special Publication 18, 644 p.

Potonié, R., and Kremp, G. O. W., 1954, Die Gattungen der paläozoischen *Sporae dispersae* und ihre Stratigraphie: Beihefte zum Geologischen Jahrbuch, v. 69 p. 111–194.

Potonié, R., and Kremp, G. O. W., 1955, Die *Sporae dispersae* des Ruhrkarbons, ihre Morphographie und Stratigraphie mit Ausblicken auf Arten anderer Gebiete und Zeitabschnitte, Teil I: Palaeontographica, v. 98, 136 p.

Potonié R., and Kremp, G. O. W., 1956, Die *Sporae dispersae* des Ruhrkarbons, ihre Morphographie und Stratigraphie mit Ausblicken auf Arten anderer Gebiete und Zeitabschnitte, Teil II: Palaeontographica, v. 99, p. 85–191.

Ramsbottom, W. H. C., ed., 1981, Field guide to the boundary stratotypes of the Carboniferous stages in Britain: Leeds, England, International Union of Geological Sciences, Subcommission on Carboniferous Stratigraphy, 105 p.

Ramsbottom, W. H. C., and Saunders, W. B., 1984, Carboniferous ammonoid zonation, *in* Sutherland, P. K., and Manger, W. L., eds., Biostratigraphy: Compte Rendu, Neuvième Congrès International de Stratigraphie et de Géologie du Carbonifère, Washington and Champaign-Urbana, 1979, v. 2, p. 52–64.

Rashid, M. A., 1968, Palynology of the Bostwick Member of the Lake Murray Formation (Pennsylvanian) of southern Oklahoma [M.S. Thesis]: Norman, University of Oklahoma, 122 p.

Ravn, R. L., 1979, An introduction to the stratigraphic palynology of the Cherokee Group (Pennsylvanian) coals of Iowa: Iowa Geological Survey Technical Paper 6, 117 p.

Ravn, R. L., 1981, Palynostratigraphy of Lower and Middle Pennsylvanian coals of Iowa, with special reference to the Atokan-Desmoinesian boundary: Geological Society of America Abstracts with Programs, v. 13, p. 314.

Ravn, R. L., 1986, Palynostratigraphy of the Lower and Middle Pennsylvanian coals of Iowa: Iowa Geological Survey Technical Paper 7, 245 p.

Ravn, R. L., and Fitzgerald, D. J., 1982, A Morrowan (Upper Carboniferous) miospore flora from eastern Iowa, U.S.A.: Paleontographica B, v. 183, p. 106–172.

Ravn, R. L., Swade, J. W., Howes, M. R., Gregory, J. L., Anderson, R. R., and Van Dorpe, P. E., 1984, Stratigraphy of the Cherokee Group and revision of Pennsylvanian stratigraphic nomenclature in Iowa: Iowa Geological Survey Technical Information Series 12, 76 p.

Ravn, R. L., Butterworth, M. A., Phillips, T. L., and Peppers, R. A., 1986, Proposed synonymy of *Granasporites* Alpern 1959 emend. and *Cappasporites* Urban emend. Chadwick 1983, miospore genera from the Carboniferous of Europe and North America: Pollen et Spores, v. 38, p. 421–433.

Read, C. B., and Mamay, S. H., 1964, Upper Paleozoic floral zones and floral provinces of the United States: U.S. Geological Survey Professional Paper 454-K, 33 p.

Remy, W., 1975, The floral changes at the Carboniferous-Permian boundary in Europe and North America, *in* Barlow, J. A., ed., The age of the Dunkard—Proceeding of the first I. C. White Memorial Symposium: Morgantown, West Virginia Geological and Economic Survey, p. 305–344.

Rexroad, C. B., and Merrill, G. K., 1979, Conodont biostratigraphy and the Mississippian-Pennsylvanian boundary in southern Illinois: Ninth International Congress of Carboniferous Stratigraphy and Geology Abstracts of Papers: Champaign-Urbana, University of Illinois, p. 179–180.

Rexroad, C. B., and Merrill, G. K., 1985, Conodont biostratigraphy and paleoecology of middle Carboniferous rocks in southern Illinois: Courier

Forschungsinstitut Senckenberg, v. 74, p. 35–64.

Rice, C. L., 1978, Ages of the Lee, Breathitt, Caseyville, Tradewater, and Sturgis formations in Kentucky, *in* Stohl, N. F., and Wright, W. B., eds., Changes in stratigraphic nomenclature by the U.S. Geological Survey, 1977: U.S. Geological Survey Bulletin 1457-A, p. 108–109.

Rice, C. L., and Smith, J. H., 1980, Correlation of coal beds, coal zones and key stratigraphic units, Pennsylvanian rocks of eastern Kentucky: U.S. Geological Survey Miscellaneous Field Studies Map MF-1188.

Rice, C. L., Kehn, T. M., and Douglass, R. C., 1979a, Pennsylvanian correlations between the Eastern Interior and Appalachian basins, *in* Palmer, J. E., and Dutcher, R. R., eds., Depositional and structural history of the Pennsylvanian System in the Illinois Basin: Part 2: Invited papers: Ninth International Congress of Carboniferous Stratigraphy and Geology, Champaign-Urbana, Field Trip 9, part 2; Illinois State Geological Survey Guidebook 15a, p. 103–105.

Rice, C. L., Sable, E. G., Dever, G. R., Jr., and Kehn, T. M., 1979b, The Mississippian and Pennsylvanian (Carboniferous) systems in the United States—Kentucky: U.S. Geological Survey Professional Paper 1110-F, 35 p.

Rogers, H. D., 1840, Fourth annual report on the geological exploration of the State of Pennsylvania: Harrisburg, Geological Survey of Pennsylvania, 215 p.

Rosowitz, D. W., 1982, Palynology and paleoecology of the Riverton Coal Bed (Desmoinesian, Pennsylvanian) in southeastern Kansas [M.S. thesis]: Wichita, Kansas, Wichita State University, 136 p.

Ross, C. A., 1984, Fusulinacean biostratigraphy near the Carboniferous-Permian boundary in North America, *in* Sutherland, P. K., and Manger, W. L., ed., Biostratigraphy: Compte Rendu, Neuvième Congrès International de Stratigraphie et de Géologie du Carbonifère, Washington and Champaign-Urbana, 1979, v. 2 p. 535–542.

Ross, C. A., and Ross, J. R. P., 1985, Late Paleozoic depositional sequences are synchronous and worldwide: Geology, v. 13, p. 194–197.

Ross, C. A., and Ross, J. R. P., 1987, Late Paleozoic sea levels and depositional sequences, *in* Ross, C. A., and Haman, D., eds., Timing and depositional history of eustatic sequences: Constraints on seismic stratigraphy: Cushman Foundation for Foraminiferal Research Special Publication 24, p. 137–149.

Ross, C. A., and Ross, J. R. P., 1988, Late Paleozoic transgressive-regressive deposition, *in* Wilgus, C. K., and others, eds., Sea-level changes: An integrated approach: Society of Economic Paleontologists and Mineralogists Special Publication no. 42, p. 227–247.

Ross, C. A., and Ross, J. R. P., 1991, Patterns in late Paleozoic sea level fluctuations and depositional sequences: Geological Society of America Annual Meeting Abstracts with Programs, v. 23, no. 5, p. A29.

Saenz de Santa Maria, J. A., Luque, C., Gervilla, M., Laveine, J. P., Loboziak, S., Brousmiche, C., Coquel, R., and Martinez-Diaz, C., 1985, Aportacion al Conocimiento estratigrafico y sedimentologico del Carbonifero productivo de la Cuenca Central Asturians, *in* Cross, A. T., ed., Economic geology: coal, oil and gas: Compte Rendu, Dixième Congrès International de Stratigraphie et de Géologie du Carbonifère, Madrid, Spain, 1983, v. 4, p. 303–326.

Saunders, W. B., and Ramsbottom, W. H. C., 1986, The mid-Carboniferous eustatic event: Geology, 14, p. 208–212.

Saunders, W. B., Manger, W. L., and Gordon, M., Jr., 1977, Upper Mississippian and Lower and Middle Pennsylvanian ammonoid biostratigraphy of northern Arkansas, *in* Sutherland, P. K., and Manger, W. L., eds., Upper Chesterian–Morrowan stratigraphy and the Mississippian-Pennsylvanian boundary in northeastern Oklahoma and northwestern Arkansas: Oklahoma Geological Survey Guidebook 18, p. 117–137.

Scheihing, M. H., and Langenheim, R. L., Jr., 1985, Depositional history of an Upper Pennsylvanian cyclothem in the Illinois Basin and comparison to Kansas cyclothemic sequences, *in* Dutro, T., Jr., and Pfefferkorn, H. W., eds., Paleontology, paleoecology, and paleogeography: Compte Rendu, Neuvième Congrès International de Stratigraphie et de Géologie du Carbonifère, Washington and Champaign-Urbana, 1979, v. 5, p. 373–382.

Schemel, M. P., 1951, Small spores of the Mystic Coal of Iowa: American Midland Naturalist, v. 46, p. 743–750.

Schemel, M. P., 1957, Small spore assemblages of mid-Pennsylvanian coals of West Virginia and adjacent areas [Ph.D. thesis]: Morgantown, West Virginia University, 190 p.

Schopf, J. M., 1936, Spores characteristics of Illinois Coal No. 6: Illinois State Academy of Science Transactions, 1935, v. 28, p. 173–176.

Schopf, J. M., 1938, Spores from the Herrin (No. 6) coal bed in Illinois: Illinois State Geological Survey Report of investigations 50, 73 p.

Schopf, J. M., Wilson, L. R., and Bentall, R., 1944, An annotated synopsis of Paleozoic fossil spores and the definition of generic groups: Illinois State Geological Survey Report of Investigations 91, 73 p.

Schutter, S. R., and Heckel, P. H., 1985, Missourian (early Late Pennsylvanian) climate in Midcontinent North America: International Journal of Coal Geology, v. 5, p. 111–140.

Scotese, C. R., 1986, Phanerozoic reconstructions: A new look at the assembly of Asia: University of Texas Institute of Geophysics Technical Report 66, 54 p.

Scotese, C. R., Bambach, R. K., Barton, C., Van der Voo, R., and Ziegler, A. M., 1979, Paleozoic base maps: Journal of Geology, v. 87, p. 217–277.

Scott, A. C., and Taylor, T. N., 1983, Plant/animal interactions during the Upper Carboniferous: Botanical Review, v. 49, p. 259–307.

Searight, W. V., and Howe, W. B., 1961, Pennsylvanian System, in Howe, W. B., and Koenig, J. W., eds., The stratigraphic succession in Missouri: Missouri Geological Survey, Second Series, v. 40, p. 78–122.

Shaver, R. H., 1984, Atokan Series concepts with special reference to the Illinois Basin and Iowa, in Sutherland, P. K., and Manger, W. L., eds., The Atokan Series (Pennsylvanian) and its boundaries—A symposium: Oklahoma Geological Survey Bulletin 136, p. 101–113.

Shaver, R. H., and Smith, S. G., 1974, Some Pennsylvanian Kirkbiacean ostracods of Indiana and Midcontinent series terminology: Indiana Geological Survey Report of Progress 31, 59 p.

Shaver, R. H., and 10 others, 1970, Compendium of rock-unit stratigraphy in Indiana: Indiana Geological Survey Bulletin 43, 229 p.

Shaver, R. H., and 31 others, 1985, Midwestern basins and arches correlation chart, sheet 8, in Childs, O. E., Correlation of stratigraphic units of North America-COSUNA: Tulsa, Oklahoma, American Association of Petroleum Geologists.

Shaver, R. H., and 16 others, 1986, Compendium of rock-unit stratigraphy in Indiana—a revision: Indiana Geological Survey Bulletin 59, 203 p.

Siever, R., 1951, The Mississippian-Pennsylvanian unconformity in southern Illinois: American Association of Petroleum Geologists Bulletin 35, p. 543–581.

Smith, A. G., Hurley, A. M., and Briden, J. C., 1981, Phanerozoic paleocontinental world maps: London, New York, Cambridge University Press, 102 p.

Smith, A. H. V., and Butterworth, M. A., 1967, Miospores in the coal seams of the Carboniferous of Great Britain: London, Palaeontology Association, Special Paper in Palaeontology, No. 1, 324 p.

Smith, A. H. V., Butterworth, M. A., Knox, E. M., and Love, L. G., 1964, Verrucosisporites (Ibrahim) emend. Report of the Commission Internationale de Microflore du Paléozoic working group no. 6: Compte Rendu, Cinquiéme Congrès International de Stratigraphie et de Géologie du Carbonifère, Paris, 1963, v. 3, p. 1071–1077.

Smyth, P., 1974, Fusulinids in the Appalachian Basin: Journal of Paleontology, v. 48, p. 856–858.

Sohn, I. G., and Jones, P. J., 1984, Carboniferous ostracodes—a biostratigraphic evaluation, in Sutherland, P. K., and Manger, W. L., eds., Biostratigraphy: Compte Rendu, Neuvième Congrès International de Stratigraphie et de Géologie du Carbonifère, Washington and Champaign-Urbana, 1979, v. 2, p. 65–80.

Solovieva, M. N., and 17 others, 1984, Biostratigraphy and correlation of the Moscovian Stage, in Gordon, J., Jr., ed., Official reports: Compte Rendu, Neuvième Congrès International de Stratigraphie et de Géologie du Carbonifère, Washington and Champaign-Urbana, 1979, v. 1, p. 83–94.

Somers, G., 1952, Fossil spore content of the Lower Jubilee seam of the Sidney coal field: Halifax, Nova Scotia Research Foundation, 30 p.

Somers, Y., 1971, Étude palynologique du Westphalien du Bassin de Campine et révision du genre Lycospora [Thèse de Doctorat]: Université de Liège.

Spivey, R. C., and Roberts, T. G., 1946, Lower Pennsylvanian terminology in central Texas: American Association of Petroleum Geologists Bulletin, v. 30, p. 181–186.

Staplin, F. L., and Jansonius, J., 1964, Elucidation of some Paleozoic densospores: Palaeontographica B, v. 114, p. 95–117.

Stepanov, D. L., and 17 others, 1962, The Carboniferous System and its main stratigraphic subdivisions; report of the Commission on the Stratigraphy of the Carboniferous of the National Committee of Soviet Geologists: Compte Rendu, Quatrième Congrès International de Stratigraphie et de Géologie du Carbonifère, Heerlen, 1958, v. 3, p. 645–656.

Stewart, W. J., 1968, The stratigraphic and phylogenetic significance of the fusulinid genus Eowaeringella with several new species: Cushman Foundation for Foraminiferal Research, Special Publication 10, 29 p.

Strimple, H. L., and Knapp, W. D., 1966, Lower Pennsylvanian fauna from eastern Kentucky, Part 2, crinoids: Journal of Paleontology, v. 40, p. 309–314.

Sullivan, H. J., 1962, Distribution of miospores through coal and shales of the Coal Measures sequence exposed in the Wernddu claypit, Caerphilly (South Wales): Quarterly Journal of Geological Society of London, v. 118, p. 353–373.

Sullivan, H. J., 1964, Miospores from the Drybrook Sandstone and associated measures in the Forest of Dean Basin, Gloucestershire: Palaeontology, v. 7, p. 351–392.

Sullivan, H. J., and Mishell, D. R., 1971, The Mississippian-Pennsylvanian boundary and its correlation with Europe: Compte Rendu, Sixième Congrès Internationale de Stratigraphie et de Géologie du Carbonifère, Sheffield, 1967, v. 4, p. 1533–1540.

Sutherland, P. K., 1979, Stop descriptions-second day, in Sutherland, P. K., and Manger, W. L., eds., Mississippian-Pennsylvanian shelf-to-basin transition Ozark and Ouachita regions, Oklahoma and Arkansas: Ninth International Congress of Carboniferous Stratigraphy and Geology, Guidebook, Field Trip 11: Oklahoma Geological Survey Guidebook 19, p. 27–37.

Sutherland, P. K., 1990, Fusulinid, conodont, and brachiopod biostratigraphy of the Morrowan and Atokan Series, Ardmore Basin, Oklahoma: Geological Society of America Abstracts with Programs, v. 22, no. 6, p. 33.

Sutherland, P. K., and Grayson, R. C., Jr., 1978, Redefinition of the Morrowan Series (Lower Pennsylvanian) in its type area in northwestern Arkansas: Geological Society of America Abstracts with Programs, v. 10, p. 501.

Sutherland, P. K., and Grayson, R. C., Jr., 1992, Morrowan and Atokan (Pennsylvanian) biostratigraphy in the Ardmore Basin, Oklahoma, in Sutherland, P. K., and Manger, W. L., eds., Recent advances in Middle Carboniferous biostratigraphy—A symposium: Oklahoma Geological Survey Circular 94, p. 81–91.

Sutherland, P. K., and Manger, W. L., 1979, Mississippian-Pennsylvanian shelf-to-basin transition Ozark and Ouachita regions, Oklahoma and Arkansas: Ninth International Congress of Carboniferous Stratigraphy and Geology, Guidebook, Field Trip 11: Oklahoma Geological Survey Guidebook 19, 81 p.

Sutherland, P. K., and Manger, W. L., 1983, The Morrowan-Atokan (Pennsylvanian) boundary problem: Geological Society of America Bulletin, v. 94, p. 543–548.

Sutherland, P. K., and Manger, W. L., 1984, An interval in search of a name, in Sutherland, P. K., and Manger, W. L., eds., The Atokan Series (Pennsylvanian) and its boundaries—A symposium: Oklahoma Geological Survey Bulletin 136, p. 1–9.

Sutherland, P. K., and Manger, W. L., eds., 1992, Recent advances in Middle Carboniferous biostratigraphy—A symposium: Oklahoma Geological Survey Circular 94, 181 p.

Swade, J. W., 1985, Conodont distribution, paleoecology, and preliminary biostratigraphy of the Upper Cherokee and Marmaton Groups (Upper Desmoinesian, Middle Pennsylvanian) from two cores in south-central Iowa: Iowa Geological Survey Technical Information Series, no. 14, 71 p.

Taff, V. A., and Adams, G. I., 1900, Geology of the eastern Choctaw coal field, Indian Territory: U.S. Geological Survey, 21st Annual Report, p. 257–311.

Tennant, S. H., 1981, Lithostratigraphy and depositional environments of the upper Dornick Hills Group (Lower Pennsylvanian) in the northern part of the Ardmore Basin, Oklahoma [M.S. thesis]: Norman, University of Oklahoma, 291 p.

Tennant, S. H., Sutherland, P. K., and Grayson, R. C., Jr., 1982, Stop description; northern Ardmore Basin, *in* Sutherland, P. K., ed., Lower Middle Pennsylvanian stratigraphy in south-central Oklahoma: Oklahoma Geological Survey Guidebook 20, p. 38–42.

Teteryuk, V. K., 1974, Palynology of the Westphalian/Stephanian boundary in the Donets Basin. Palynology of the proterophytic and paleophytic: Proceedings of the 3rd International Palynological Conference: Nauka (Moscow, USSR) (Trudy III Mezhdunarodnvi palinologicheskoi konferentsii). Translated by Canadian Multilingual Services Division, p. 114–116.

Teteryuk, V. K., 1976, Namurian stage analogues in the Carboniferous Period of the Donetz Basin (based on palynological data): Geologicheshii Zhurnal, v. 36, p. 110–122.

Teteryuk, V. K., 1982, On the Carboniferous Bashkirian-Moscovian boundary of the east European platform, *in* Ramsbottom, W. H. C., Saunders, W. B., and Owens, B., eds., Biostratigraphic data for a Mid-Carboniferous boundary: Leeds, England, International Union of Geological Sciences Subcommission on Carboniferous Stratigraphy, Institute of Geological Sciences, p. 36–41.

Thompson, M. L., 1934, The fusulinids of the Des Moines Series of Iowa: Iowa University Studies in Natural History, v. 16, p. 273–332.

Thompson, M. L., 1936, Pennsylvanian fusulinids from Ohio: Journal of Paleontology, v. 10, p. 673–682.

Thompson, M. L., 1942, Pennsylvanian System in New Mexico: New Mexico School of Mines Bulletin 17, 90 p.

Thompson, M. L., 1948, Protozoa: Studies of American fusulinids: University of Kansas Paleontological Contribution 14, 184 p.

Thompson, M. L., and Riggs, E. A., 1959, Part 1—Primitive fusulinids in the southern part of the Illinois Basin: Journal of Paleontology, v. 33, p. 771–781.

Thompson, M. L., Shaver, R. H., and Riggs, E. A., 1959, Early Pennsylvanian fusulinids and ostracods of the Illinois Basin: Journal of Paleontology, v. 33, p. 770–792.

Thompson, T. L., 1979, The Mississippian and Pennsylvanian (Carboniferous) systems in the United States—Missouri: U.S. Geological Survey Professional Paper 1110-N, 22 p.

Trask, C. B., and Jacobson, R. J., 1990, Geologic map of the Creal Springs quadrangle, Illinois: Illinois State Geological Survey Map IGQ-4, scale 1:24,000.

Udden, J. A., 1912, Geology and mineral resources of the Peoria quadrangle, Illinois: U.S. Geological Survey Bulletin 506, p. 1–103.

Unklesbay, A. G., 1954, Distribution of American Pennsylvanian cephalopods: Journal of Paleontology, v. 28, p. 84–95.

Unuigboje, F. E., 1987, Palynology of the Fire Clay coal bed of eastern Kentucky [M.S. thesis]: Richmond, Eastern Kentucky University, 59 p.

Upshaw, C. F., and Hedlund, R. W., 1967, Microspores from the upper part of the Coffeyville Formation (Pennsylvanian, Missourian), Tulsa County, Oklahoma: Pollen et Spores, v. 9, p. 143–170.

Urban, L. L., 1965, Palynology of the Drywood and Bluejacket coals (Pennsylvanian) of Oklahoma [M.S. thesis]: Norman, University of Oklahoma, 91 p.

Vail, P. R., Mitchum, R. M., Jr., and Thompson, S., III, 1977, Seismic stratigraphy and global changes of sea level, Part 4: Global cycles of relative changes of sea level, *in* Payton, C. E., ed., Seismic stratigraphy—applications to hydrocarbons exploration: American Association of Petroleum Geologists Memoir 26, p. 83–97.

Van de Laar, J. G. M., and Fermont, W. J. J., 1989, On-shore Carboniferous palynology of The Netherlands: Mededelingen van's Rijks Geologische dienst, v. 43, p. 36–68.

Van de Laar, J. G. M., and Fermont, W. J. J., 1990, Westphalian palynology of The Netherlands based on six continuously cored boreholes: Review of Paleobotany and Palynology, v. 65, p. 275–285.

van Ginkel, A. C., 1965, Carboniferous fusulinids from the Cantabrian Mountains: Leidsche Geologische Mededelingen, v. 34, 225 p.

van Ginkel, A. C., 1972. Correlation of the Myachkovian and Kasimovian in the USSR with European subdivision: Leidsche Geologische Mededelingen, v. 49, p. 1–7.

Van Wijhe, D. H., and Bless, M. J. M., 1974, The Westphalian of The Netherlands with special reference to miospore assemblages: Geologie en Mijnbouw, v. 53, p. 295–328.

Von Bitter, P. H., and Merrill, G. K., 1980, Naked species of *Gondolella* (Conodontophorida): Their distribution, taxonomy, and evolutionary significance: Royal Ontario Museum, Life Sciences Contribution 125, 49 p.

Wagner, R. H., 1964, Stephanian floras in N. W. Spain with special reference to the Westphalian D-Stephanian A boundary: Compte Rendu, Cinquième Congrès International de Stratigraphie et de Géologie du Carbonifère, Paris, 1963, v. 2, p. 835–851.

Wagner, R. H., 1966, El significado de la flora en la estratigrafia del Carbonifero superior: Boletin de la Real Sociedad Espanola de Historia Natural, v. 64, p. 203–208.

Wagner, R. H., 1969, Proposal for the recognition of a new "Cantabrian" Stage at the base of the Stephanian Series: Compte Rendu, Sixième Congrès International de Stratigraphie et de Géologie du Carbonifère, Sheffield, 1967, v. 1, p. 139–150.

Wagner, R. H., 1976, The chronostratigraphic units of the Upper Carboniferous in Europe: Société Géologique de Belgique Bulletin, v. 83, p. 235–253.

Wagner, R. H., 1984, Megafloral zones of the Carboniferous, *in* Sutherland, P. K., and Manger, W. L.., eds., Biostratigraphy: Compte Rendu, Neuvième Congrès International de Stratigraphie et de Géologie du Carbonifère, Washington and Champaign-Urbana, 1979, v. 2, p. 109–134.

Wagner, R. H., 1989, A late Stephanian forest swamp with *Sporangiostrobus* fossilized by volcanic ash fall in the Puertollano Basin, central Spain: International Journal of Coal Geology, v. 12, p. 523–552.

Wagner, R. H., and Bowman, M. B. J., 1983, The position of the Bashkirian-Moscovian boundary in West European chronostratigraphy: Newsletters on Stratigraphy, v. 12, p. 132–161.

Wagner, R. H., and Higgins, A. C., 1979, The Carboniferous of the USSR: its stratigraphic significance and outstanding problems of world-wide correlation, *in* Wagner, R. H., Higgins, A. C., and Meyen, S. V., eds., Compte Rendu, Huitième International Congrès Stratigraphie et de Géologie du Carbonifère, Moscow, 1975, p. 5–22.

Wagner, R. H., and Spinner, E., 1976, *Bodeodendron*, troc associé à *Sporangiostrobus*: Paris, Academie des Sciences, Comptes Rendus, sér. D, v. 282, p. 353–356.

Wagner, R. H., and Varker, W. J., 1971, The distribution and development of post-Leonian strata (upper Westphalian D, Cantabrian, Stephanian A) in northern Palencia, Spain: Trabajos de Geologia, Oviedo, v. 4, p. 533–601.

Wagner, R. H., and Winkler Prins, C. F., 1979, The lower Stephanian of western Europe, *in* Meyen, S. V., ed., Palaeontological characteristics of the main subdivisions of the Carboniferous: Compte Rendu, Huitième Congrès International de Stratigraphie et de Géologie du Carbonifère, Moscow, 1975, v. 3, p. 111–140.

Wagner, R. H., and Winkler Prins, C. F., 1985, Stratotypes of the two lower Stephanian stages, Cantabrian and Barruelian: Compte Rendu, Dixième Congrès International Stratigraphie et de Géologie du Carbonifère: Madrid, Spain, 1983, v. 4, p. 473–483.

Wagner, R. H., Spinner, E., Jones, D. G., and Wagner-Gentis, C. H. T., 1970, The upper Cantabrian rocks near Inguanzo, eastern Asturias, Spain: Colloque sur la stratigraphie du Carbonifère: Les Congres et Colloques de l' Université de Liège, v. 55, p. 465–486.

Wagner, R. H., Park, R. K., Winkler Prins, C. F., and Lys, M., 1977, The Post-Leonian Basin in Palencia, a report on the stratotype of the Cantabrian Stage: Symposium on Carboniferous Stratigraphy: Geological Survey of Prague, p. 89–146.

Wanless, H. R., 1931, Pennsylvanian cycles in western Illinois, in Papers presented at the quarter centennial celebration of the Illinois State Geological Survey: Illinois State Geological Survey Bulletin, v. 60, p. 179–193.

Wanless, H. R., 1939, Pennsylvanian correlations in the Eastern Interior and Appalachian coal fields: Geological Society of America Special Paper 17, 130 p.

Wanless, H. R., 1956, Classification of the Pennsylvanian rocks of Illinois as of 1956: Illinois State Geological Survey Circular 217, 14 p.

Wanless, H. R., 1957, Geology and mineral resources of the Beardstown, Glasford, Havana, and Vermont quadrangles: Illinois State Geological Survey Bulletin 82, 233 p.

Wanless, H. R., 1958, Pennsylvanian faunas of the Beardstown, Glasford, Havana, and Vermont quadrangles: Illinois State Geological Survey Report of Investigations 205, 59 p.

Wanless, H. R., 1975, Illinois Basin region, in McKee, E. D., and others, eds., Paleotectonic investigations of the Pennsylvanian System in the United States: U.S. Geological Survey Professional Paper 853, part 1, p. 71–95.

Wanless, H. R., and Weller, J. M., 1932, Correlation and extent of Pennsylvanian cyclothems: Geological Society of America Bulletin, v. 43, p. 1003–1016.

Weibel, C. P., 1986, Lithostratigraphic revision within Upper Pennsylvanian Mattoon Formation, Illinois: American Association of Petroleum Geologists Bulletin, v. 69, p. 315.

Weibel, C. P., and Norby, R. D., 1990, Sedimentologic and micropaleontologic study of a core through the Mississippian-Pennsylvanian boundary in southern Illinois: Geological Society of America Abstracts with Programs, v. 11, p. 35.

Weibel, C. P., Nelson, W. J., Oliver, L. B., and Esling, S. P., 1993, Geology of the Waltersburg quadrangle, Pope County, Illinois: Illinois State Geological Survey Bulletin, v. 98, 41 p.

Weller, J. M., 1930, Cyclical sedimentation of the Pennsylvanian period and its significance: Journal of Geology, v. 38, p. 97–135.

Weller, J. M., 1931, The conception of cyclical sedimentation during the Pennsylvanian period; in Papers presented at the quarter centennial celebration of the Illinois State Geological Survey: Illinois State Geological Survey Bulletin, v. 60, p. 163–179.

Weller, J. M., 1964, Development of the concept and interpretation of cyclic sedimentation, in Merriam, D. F., ed., Symposium on cyclic sedimentation: Kansas Geological Survey Bulletin 169, p. 604–621.

Weller, S., 1906, The geologic map of Illinois: Illinois State Geological Survey Bulletin, v. 1, 24 p.

West, R. R., and Archer, A. W., 1992, Upper Carboniferous reef mounds and climate change: Geological Society of America Abstracts with Programs, v. 24, no. 7, p. 119.

White, D., 1896, Report on the fossil plants from the Hindostan Whetstone beds in Orange County, Indiana, in Kindle, E. M., ed., The whetstone and grindstone rocks of Indiana: Indiana Department Geology and Natural Resources Annual Report, v. 20, p. 354–355.

White, D., 1900, The stratigraphic succession of the fossil floras of the Pottsville formation in the southern anthracite coal field, Pennsylvania: U.S. Geological Survey 20th Annual Report, part 2, p. 751–930.

White, I. C., 1908, Supplementary coal report: West Virginia Geological Survey, Volume 2: Morgantown, West Virginia, 720 p.

Wilgus, C. K., Hastings, B. S., Kendall, C. G. St. C., Posamentier, H. W., Ross, C. A., and Van Wagoner, J. C., eds., 1988, Sea-level changes: An integrated approach: Society of Economic Paleontologists and Mineralogists Special Publication 42, 407 p.

Williams, D. A., Williamson, A. D., and Beard, J. G., 1982, Stratigraphic framework of coal-bearing rocks in the Western Kentucky Coal Field: Kentucky Geological Survey, Series XI, Information Circular 8, 201 p.

Williams, R. W., 1955, Pityosporites westphalensis sp. nov., an abietineous type

pollen grain from the Coal Measures of Britain: Annals and Magazine of Natural History, v. 8, p. 465–473.

Williamson, A. D., and McGrain, P., 1979, Rough River State Resort Park to Kentucky Dam Village State Resort Park, in Palmer, J. F., and Dutcher, R. R., eds., Depositional and structural history of the Illinois Basin: Part 1: Road Log and description of stops: Ninth International Congress of Carboniferous Stratigraphy and Geology, Champaign-Urbana, Part I, Field Trip 9, part 1; Illinois State Geological Survey Guidebook Series 15a, p. 37–53.

Willman, H. B., and 9 others, 1967, Geologic map of Illinois: Illinois State Geological Survey, scale 1:500,000.

Wilson, L. R., 1965, Palynological age determination of a rock section in Ti Valley, Pittsburg County, Oklahoma: Oklahoma Geological Notes, v. 25, p. 11–18.

Wilson, L. R., 1970, Palynology of Oklahoma's ten-foot coal seam: Oklahoma Geological Notes, v. 30, p. 62–63.

Wilson, L. R., 1972, Fossil plants of the Seminole Formation (Pennsylvanian) in Tulsa County, Oklahoma: Tulsa Geological Society Digest, v. 37, p. 151–161.

Wilson, L. R., 1976a, Desmoinesian coal seams of northeastern Oklahoma and their palynological content: Tulsa Geological Society, Field Trip Guidebook, p. 19–32.

Wilson, L. R., 1976b, A preliminary assessment of palynomorph stratigraphy in Morrow and Atoka rocks of the Ouachita Mountains, Oklahoma. A study of Paleozoic rocks in Arbuckle and western Ouachita mountains of southern Oklahoma: Gulf Coast Association of Geological Societies, Field Trip Guidebook, p. 83–89.

Wilson, L. R., 1979a, Palynologic and plant compression evidence for Desmoinesian-Missourian (Pennsylvanian) Series boundary in northeastern Oklahoma: American Association of Petroleum Geologists Bulletin, v. 63, p. 2120.

Wilson, L. R., 1979b, Palynological and plant compression evidence for a Desmoinesian- Missourian (Pennsylvanian) Series boundary in northeastern Oklahoma: Ninth International Congress of Carboniferous Stratigraphy and Geology, Abstract of Papers, Champaign-Urbana, p. 234–235.

Wilson, L. R., 1979c, Palynological evidence for age assignment of basal Cherokee (Pennsylvanian) strata in southeastern Kansas: Oklahoma Geology Notes, v. 39, p. 76–79.

Wilson, L. R., 1984, Evidence for a new Desmoinesian-Missourian boundary (Middle Pennsylvanian) in Tulsa County, Oklahoma, U.S.A., in Sharma, A. K., Mitra, G. C., and Banerjee, M., eds., A. K. Ghosh, Commemorative Volume: New Delhi, Today and Tomorrow Printers and Publishers, p. 251–265.

Wilson, L. R., and Bennison, A. P., 1981, Multidiscipline evidence for a new Desmoinesian-Missourian (Pennsylvanian) boundary in central Oklahoma: Palynology, v. 5, p. 244–245.

Wilson, L. R., and Coe, E. A., 1940, Description of some unassigned plant microfossils from the Des Moines Series of Iowa: American Midland Naturalist, v. 23, p. 182–186.

Wilson, L. R., and Hoffmeister, W. S., 1956, Plant microfossils of the Croweburg Coal: Oklahoma Geological Survey Circular 32, 57 p.

Wilson, L. R., and Kosanke, R. M., 1944, Seven new species of unassigned plant microfossils from the Des Moines series of Iowa: Iowa Academy of Science Proceedings, v. 51, p. 329–332.

Wilson, L. R., and Rashid, M. A., 1982, Paleobotanical evidence for an age assignment of the Bostwick Member (Pennsylvanian): Geological Society of America Abstracts with Programs, v. 14, p. 140.

Wilson, L. R., and Venkatachala, B. S., 1963a, Thymospora, a new name for Verrucososporites: Oklahoma Geological Notes, v. 23, p. 75–79.

Wilson, L. R., and Venkatachala, B. S., 1963b, An emendation of Vestispora Wilson and Hoffmeister, 1956: Oklahoma Geology Notes, v. 23, p. 94–100.

Wilson, L. R., and Venkatachala, B. S., 1963c, A morphologic study and emendation of Vesicaspora Schemel, 1951: Oklahoma Geology Notes, v. 23, p. 142–149.

Wilson, L. R., and Venkatachala, B. S., 1964, *Potonieisporites elegans* (Wilson and Kosanke) comb. nov.: Oklahoma Geological Notes, v. 24, p. 67–68.

Wilson, L. R., and Venkatachala, B. S., 1968, Palynological composition and succession in a Dawson coal (Pennsylvanian) section of Oklahoma: Geological Society of America Special Paper 115, p. 381.

Winslow, M. R., 1959, Upper Mississippian and Pennsylvanian megaspores and other plant microfossils from Illinois: Illinois Geological Survey Bulletin 86, 135 p.

Worthen, A. H., 1875, Geology and paleontology: Illinois State Geological Survey, v. 6, 532 p.

Zachry, D. L., and Sutherland, P. K., 1984, Stratigraphy and depositional framework of the Atoka Formation (Pennsylvanian) Arkoma Basin of Arkansas and Oklahoma, *in* Sutherland, P. K., and Manger, W. L., eds., The Atokan Series (Pennsylvanian) and its boundaries—A symposium: Oklahoma Geological Survey Bulletin 136, p. 9–17.

Zeller, D. E., Jewett, J. M., Bayne, C. K., Goebel, E. D., O'Connor, H. G., and Swineford, A., 1968, The stratigraphic succession in Kansas: Geological Survey of Kansas Bulletin 189, 81 p.

Ziegler, A. M., Bambach, R. K., Parrish, J. T., Barrett, S. F., Gierlowski, E. H., Parker, W. C., Raymond, A., and Sepkoski, J. J., Jr., 1981, Paleozoic biogeography and climatology, *in* Niklas, K. J., ed., Paleobotany, paleoecology, and evolution: New York, Praeger, p. 231–266.

Ziegler, A. M., Scotese, C. R., McKerrow, W. S., Johnson, M. E., and Bambach, R. K., 1979, Paleozoic paleogeography: Annual Review of Earth and Planetary Science, v. 7, p. 473–502.

Zimbrick, G. D., 1978, The lithostratigraphy of the Morrowan Bloyd and McCully formations (Lower Pennsylvanian) in southern Adair and parts of Cherokee and Sequoyah counties, Oklahoma [M.S. thesis]: Norman, University of Oklahoma, 182 p.

Zodrow, E. L., 1986, Succession of paleobotanical events: evidence for mid-Westphalian D floral changes, Morien Group (Late Pennsylvanian) Nova Scotia: Review of Palaeobotony and Palynology, v. 47, p. 293–326.

MANUSCRIPT ACCEPTED BY THE SOCIETY MAY 26, 1995

Index

[Italic page numbers indicate major references]

Typeset in U.S.A. by Johnson Printing, Boulder, Colorado
Printed in U.S.A. by Malloy Lithographing, Inc., Ann Arbor, Michigan